Natural Products Chemistry of Botanical Medicines from Cameroonian Plants

Natural Products Chemistry of Global Plants

Editor

Raymond Cooper

This unique book series focuses on the natural products chemistry of botanical medicines from different countries such as Sri Lanka, Cambodia, Brazil, China, Africa, Borneo, Thailand, and Silk Road countries. These fascinating volumes are written by experts from their respective countries. The series will focus on the pharmacognosy, covering recognized areas rich in folklore as well as botanical medicinal uses as a platform to present the natural products and organic chemistry. Where possible, the authors will link these molecules to pharmacological modes of action. The series intends to trace a route through history from ancient civilizations to the modern day showing the importance to man of natural products in medicines, foods, and a variety of other ways.

RECENT TITLES IN THIS SERIES:

Natural Products Chemistry of Botanical Medicines
from Cameroonian Plants
Xavier Siwe Noundou

Medicinal Plants and Mushrooms of Yunnan Province of China
Clara Lau and Chun-Lin Long

Medicinal Plants of Borneo
Simon Gibbons and Stephen P. Teo

Natural Products and Botanical Medicines of Iran
Reza Eddin Owfi

Natural Products of Silk Road Plants
Raymond Cooper and Jeffrey John Deakin

Brazilian Medicinal Plants
Luzia Modolo and Mary Ann Foglio

Medicinal Plants of Bangladesh and West Bengal: Botany,
Natural Products, and Ethnopharmacology
Christophe Wiart

Traditional Herbal Remedies of Sri Lanka
Viduranga Y. Waisundara

Natural Products Chemistry of Botanical Medicines from Cameroonian Plants

Edited by
Xavier Siwe-Noundou

CRC Press
Taylor & Francis Group
Boca Raton London New York

CRC Press is an imprint of the
Taylor & Francis Group, an **informa** business

First edition published 2022
by CRC Press
6000 Broken Sound Parkway NW, Suite 300, Boca Raton, FL 33487-2742

and by CRC Press
2 Park Square, Milton Park, Abingdon, Oxon, OX14 4RN

© 2022 Taylor & Francis Group, LLC

CRC Press is an imprint of Taylor & Francis Group, LLC

Library of Congress Cataloging-in-Publication Data
Names: Siwe-Noundou, Xavier, editor.
Title: Natural products chemistry of botanical medicines from Cameroonian plants /
   [edited by] Xavier Siwe-Noundou.
Description: First edition. | Boca Raton : CRC Press, 2022. | Series: Natural products chemistry
   of global plants | Includes bibliographical references and index.
Identifiers: LCCN 2021015104 (print) | LCCN 2021015105 (ebook) | ISBN
   9781032064765 (hardback) | ISBN 9781138581425 (paperback) | ISBN 9780429506734 (ebook)
Subjects: LCSH: Medicinal plants—Cameroon. | Materia medica, Vegetable—Cameroon. |
   Pharmacognosy—Cameroon. | Pharmaceutical chemistry—Cameroon. |
   Botanical chemistry—Cameroon.
Classification: LCC RS180.C26 N38 2022 (print) | LCC RS180.C26 (ebook) |
   DDC 615.9/45096711—dc23
LC record available at https://lccn.loc.gov/2021015104
LC ebook record available at https://lccn.loc.gov/2021015105

ISBN: 9781032064765 (hbk)
ISBN: 9781138581425 (pbk)
ISBN: 9780429506734 (ebk)

DOI: 10.1201/9780429506734

Typeset in Times
by codeMantra

Contents

Book Series: Natural Products Chemistry of Global Plants

CRC Press is publishing a new book series on the *Natural Products Chemistry of Global Plants*. This new series focuses on pharmacognosy, covering recognized areas rich in folklore and botanical medicinal uses as a platform to present the natural products and organic chemistry and where possible, link these molecules to pharmacological modes of action. This book series on the botanical medicines from different countries includes, but is not limited to, Bangladesh, Borneo, Brazil, Cambodia, Cameroon, Carpathia, Ecuador, Iran, Laos, Madagascar, South Africa, Sri Lanka, The Silk Road, Thailand, Trans-Himalaya, Turkey, Uganda, Vietnam, and Yunnan Province (China) and is written by experts from each country. The intention is to provide a platform to bring forward information from under-represented regions.

Medicinal plants are an important part of human history, culture, and tradition. Plants have been used for medicinal purposes for thousands of years. Anecdotal and traditional wisdom concerning the use of botanical compounds is documented in the rich histories of traditional medicines. Many medicinal plants, spices, and perfumes changed the world through their impact on civilization, trade, and conquest. Folk medicine is commonly characterized by the application of simple indigenous remedies. People who use traditional remedies may not understand in our terms the scientific rationale for why they work but know from personal experience that some plants can be highly effective.

This series provides rich sources of information from each region. An intention of the series of books is to trace a route through history from ancient civilizations to the modern day showing the important value to humankind of natural products in medicines, in foods, and in many other ways. Many of the extracts are today associated with important drugs, nutrition products, beverages, perfumes, cosmetics, and pigments, which will be highlighted.

The books are written for both chemistry students who are at university level and for scholars wishing to broaden their knowledge in pharmacognosy. Through examples of chosen herbs and plants, the series describes the key natural products and their extracts with emphasis upon sources, an appreciation of these complex molecules and applications in science.

In this series, the chemistry and structure of many substances from each region will be presented and explored. Often, books describing folklore medicine do not describe the rich chemistry or the complexity of the natural products and their respective biosynthetic building blocks. The story becomes more fascinating by drawing upon the chemistry of functional groups to show how they influence the chemical behavior of the building blocks which make up large and complex natural products. Where possible, it will be advantageous to describe the pharmacological nature of these natural products.

R Cooper Ph.D., Editor-in-Chief
Dept. Applied Biology & Chemical Technology
The Hong Kong Polytechnic University
Hong Kongrcooperphd@aol.com

Introduction – Cameroon

Located in Central Africa, the Republic of Cameroon (Figure A) stretches from the Gulf of Guinea to Lake Chad, between the 2nd and 13th degree of North latitude and the 9th and 16th degree of east longitude. It is limited to the north by Lake Chad, to the west by Nigeria and the Atlantic Ocean, to the east by the Central African Republic and Chad, and to the south by Equatorial Guinea, Congo, Gabon. It has an area of 475,442 km². Cameroon stands out in Africa by its physical diversity (varied relief, varied range of vegetation, several types of soil, climatic diversity, etc.). All these elements and its history make the country an "Africa in miniature."

Cameroon is in the intertropical zone and has two main types of climate: the tropical climate in the north and the equatorial climate in the south. The tropical climate extends from Adamaoua region to Lake Chad and has two climatic variations: the tropical Sahelian climate with irregular rains and accentuated drought and the Sudanese tropical climate characterized by a long rainy season (6–7 months) and a dry season of about 5 months. The equatorial climate is characterized by the abundance and regularity of rainfall but has two variations: the Guinean equatorial climate that reigns over the entire South Cameroonian plateau and the Cameroonian equatorial climate that reigns in the highlands of the West and the coast.

The vegetation adapted to the tropical climate is a savannah with multiple aspects: wooded savannah and open Sudanese forest; thorny steppe in the Sahelian. In the equatorial domain, there is the dense forest on the southern Cameroonian plateau, the mangroves with mangroves in the coast and the wooded savannah on the highlands of the west.

The rainforest in Africa is wide, diverse, and rich in important medicinal plants used by the local population to manage and treat several diseases. The local populations are poor and in this regard cannot afford to secure modern medicine to combat diseases. Cameroon constitutes an important part of Africa rainforest. A high number of known medicinal plants are used by local populations as traditional medicine to treat various diseases including malaria, diabetes, cancer, inflammations, viruses, and microbial infections.

In rural, but also increasingly in urban areas of Cameroon, the practice of traditional medicine is common and frequent. It is obvious to say that most of the population living there has already experimented with traditional medicine. In fact, when there is a case of sickness, the first instinct is to resort to traditional medicine and mainly to the use of medicinal plants and natural substances for any treatment. When the patients find relief or a cure, they tend to forget modern medicine and therefore do not go to the hospital. Most Cameroonian rural areas are rich in fauna and flora biodiversity. There is a wide variety of medicinal plants (ranging from herbs to higher plants) with various and varied therapeutic virtues. In general, traditional healers offer potions, powders, or decoctions that help to manage different diseases, after consultations. The practice of traditional medicine in general is transmitted from parents to children, but recently, the general observation is that there is loss or breakage of transfer of traditional knowledge, due either to the lack of interests of the new generation of population or because of the parents who do not invest in passing the knowledge on to their offspring but above all because of the absence of written archives, which can immortalize knowledge and operating methods in traditional medicine. One of the main objectives of this book on medicinal plants of Cameroon is therefore to contribute to the conservation of knowledge of traditional African medicine in general and to the enhancement and archiving of knowledge based on the use of plants in traditional Cameroonian medicine in particular.

Xavier Siwe-Noundou Ph.D., Editor
Department of Chemistry
Rhodes University
South Africa
x.siwenoundou@ru.ac.za

FIGURE A   Map of Cameroon with its ten regions' delimitations.

Editor

Dr. Xavier Siwe-Noundou is a scholar and scientist based at Rhodes University in Grahamstown / Makhanda, South Africa. He has been a Gordon and Betty Moore Foundation (GBMF) Fellow (2017-2019), Oregon State University in USA (2017), University of California San Diego in USA (2019); an EU FP7 Marie Curie Fellow (2015–2016), Kaposvar University in Hungary (2015, 2016), Trakia Univesity in Bulgaria (2016), TWAS Fellow (2013), National Research Foundation South Africa Fellow (2014-2016).

Dr. Siwe-Noundou works on medicinal chemistry focusing on chemistry, pharmacognosy, and nanotechnology. His main investigations include terrestrial natural products chemistry (from Cameroon and South Africa) and marine natural products chemistry (from South Africa cost line). All the bioactive metabolites isolated are screened as potential antimicrobial, antiparasitic and antiproliferative candidates. He is an author of more than fifty scientific publications in his field of expertise.

Contributors

Matthias Onyebuchi Agbo
Department of Pharmaceutical
and Medicinal Chemistry
University of Nigeria
Nsukka, Nigeria

Léon Azefack Tapondjou
Department of Chemistry
University of Dschang
Dschang, Cameroon

Luciano Barboni
School of Science and Technology,
Chemistry Division
University of Camerino
Camerino, Italy

Gabin T. M. Bitchagno
Department of Chemistry
University of Dschang
Dschang, Cameroon

Bazil Ekuh Ewane
Department of Pharmaceutical
and Medicinal Chemistry
Faculty of Pharmaceutical Science

Serge Alain Fobofou
Department of Biological Chemistry
and Molecular Pharmacology
Harvard Medical School
Boston, Massachusetts
Technische Universität
Braunschweig
Braunschweig, Germany

Eleonora D. Goosen
Division of Pharmaceutical
Chemistry
Faculty of Pharmacy
Rhodes University
Makhanda, South Africa

Beaudelaire Kemvoufo Ponou
Department of Chemistry
University of Dschang
Dschang, Cameroon

Tanyi M. Kima
Division of Pharmaceutical Chemistry
Faculty of Pharmacy
Rhodes University
Makhanda, South Africa

Rui W.M. Krause
Department of Chemistry
Rhodes University
Makhanda, South Africa

Ezealisiji Kenneth Maduabuchi
Department of Pharmaceutical
and Medicinal Chemistry
University of Port Harcourt
Port Harcourt, Nigeria

Vaderament-A Nchiozem-Ngnitedem
Department of Chemistry
University of Nairobi
Nairobi, Kenya

Xavier Siwe-Noundou
Department of Chemistry
Rhodes University
Makhanda, South Africa

Patience O. Osadebe
Department of Pharmaceutical
and Medicinal Chemistry
University of Nigeria
Nsukka, Nigeria

Benjamin C. Ozumba
Department of Obstectrics
University of Nigeria
Nsukka, Nigeria

Sylvain Valere Sob
Department of Pharmaceutical Sciences
School of Pharmacy
University of Maryland
Baltimore, Maryland

Jean Emmanuel Mbosso Teinkela
Department of Biological Sciences
University of Douala
Douala, Cameroon
RD3 Department
Université Libre de Bruxelles (ULB)
Bruxelles, Belgium

Romuald Tematio Fouedjou
Department of Chemistry
University of Dschang
Dschang, Cameroon

Rémy Bertrand Teponno
Department of Chemistry
University of Dschang
Dschang, Cameroon

Billy Tchegnitegni Toussie
Department of Chemistry
University of Dschang
Dschang, Cameroon

Bienvenu Tsakem
Department of Chemistry
University of Dschang
Dschang, Cameroon

Philip F. Uzor
Department of Pharmaceutical
    and Medicinal Chemistry
Faculty of Pharmaceutical Science
University of Nigeria
Nsukka, Nigeria

1 The Genus *Olax*: Traditional Uses, Phytochemistry and Biological Activities

Bienvenu Tsakem, Beaudelaire Kemvoufo Ponou, Billy Tchegnitegni Toussie, and Romuald Fouedjou Tematio
University of Dschang

Xavier Siwe Noundou and Rui W.M. Krause
Rhodes University

Rémy Bertrand Teponno and Léon Azefack Tapondjou
University of Dschang

CONTENTS

DOI: 10.1201/9780429506734-1

1.1   INTRODUCTION

Microbial diseases are considered as a challenge to public health because of the unavailability of vaccines or limited chemotherapy [1]. As highlighted by the recent World Health Organization (WHO) report in 2019, the most critical group includes multi-resistant bacteria that pose a particular threat in hospitals and nursing home and among patients [2]. Certain chemical species such as free radicals are the main cause of many diseases such as Alzheimer's disease and diabetes [3]. In addition, malaria, tuberculosis and AIDS are implicated in the extraordinary morbidity and mortality in Sub-Saharan Africa [4]. From ancient time, the world population has considered plants as their sources of drugs to treat and manage diseases [5]. Almost 80% of the world population relies on traditional medicine for their primary healthcare according to the WHO report [6], and this is one of the reasons why many researchers continue to focus their studies on medicinal plants. Particularly in several regions of rural Africa, traditional healers prescribing medicinal plants find that they are the most easily accessible and affordable health resource available to the local community [7].

The *Olax* genus is a group of plants belonging to the family Olacaceae and distributed throughout Central and West Africa and Asia. It comprises about 200 species [8]. This genus has been employed for a long time in Cameroon, Nigeria, Ivory Coast and India as a herbal drug. The commonly used species include *Olax subscorpioidea*, *O. mannii* and *O. scandens*. Plant species of *Olax* are largely used for treatment of yellow fever, jaundice, mouth ulcer and malaria [9,10]. A survey conducted by Beche in the Centre Region of Cameroon during the Nachtigal project led to the discovery of local use of *Olax subscorpioidea* (locally known as "*Olom bekoé*") by Eton as spices and a remedy used to prevent many childhood illnesses [11]. Several phytochemical investigations led to the isolation of saponins and flavonoids [35–39], and an important number of pharmacological studies have shown that these plants possess antimicrobial, anticholinesterase and hypoglycemic activities [10,12,45,47,52–54]. This chapter focuses on the traditional uses and phytochemical and pharmacological investigations of the genus *Olax* over the past few decades.

1.2 TRADITIONAL USES OF PLANT SPECIES FROM *OLAX* GENUS

Considered as the oldest in the world, African traditional medicine is widespread in many localities and provides communities with enough plants for their healthcare [7]. Traditional healers deliver secrets about traditional uses of the plants depending on the locality. However, through surveys, researchers have gathered substantial amounts of information about the traditional uses of plants of the genus *Olax* (Table 1.1).

1.2.1 *OLAX MANNII*

This plant is widely found in the tropics especially in Nigeria, Ghana and Sierra Leone. It is a climbing shrub that grows up to 2 m high. It is largely used in Nigeria by "Ibo" people; the local Ibo name is *"Ngborogwu arua"*, and Hausa people call it *"Tsada biri"*. Decoctions of the leaves and roots of this plant are used for the treatment of fever, yellow fever and snake bites. A foamy solution resulting when the root bark of *Olax mannii* is soaked in water and shaken thoroughly serves as a potent anticonvulsant agent in Ohafia in Abia State, Nigeria [12]. The twigs are used as chewing sticks in Ghana [13]. Hausa tribes of Kaduna State (Nigeria) use the roots of *Olax mannii* as an antidepressant: the preparation consists of an infusion of roots in water, and a half-cup is taken orally; pure honey can also be added by mixing it with root powder for the same purpose [14].

1.2.2 *OLAX SUBSCORPIOIDEA*

Olax subscorpioidea Oliv is a shrub or tree (Figure 1.1) widely distributed in West and Central African countries such as Nigeria, Cameroon, Zaire and Senegal. It is used in folk medicine mostly for giving relief for several ailments. "Eton" people from the central region of Cameroon use this plant as spices and to treat childhood illnesses; the plant is known as *"Olom bekoé"*. Fruits of this plant are sold in West Cameroon as spices and have been recently shown to be a potential anticandidal remedy [15]. Locally called *"Eza-bor"* by Ogba/Egbema/Ndoni ethnics of Rivers State in Nigeria, roots of *O. subscorpioidea* are used to treat irregular menstrual periods. The preparation consists of extraction of its roots with dry gin and taking one shot twice daily [16]. Also, in Ogun State of Nigeria, its roots are used for the treatment of cancer [17]. Ogoni people (Southeastern Senatorial District in River State Nigeria) use the decoction of leaves, stem and roots of *O. subscorpioidea* to heal jaundice and yellow fever. Infusion of its roots is used in convulsion conditions, and a chewing stick helps alleviate toothache in the same locality [18]. Also, in Nigeria, its leaves are traditionally used in the treatment of diabetes mellitus, and they either form part of a traditional recipe or are used singly in diabetic conditions. *O. subscorpioidea* is a major component of a recipe used traditionally in the management of obesity. Of note, a survey was undertaken in three herbal markets in Lagos State, Nigeria, to obtain information on local herbal remedies used for the treatment of dental/oral disorders, and the results show that the stems/roots of *O. subscorpioidea* are used for this purpose by Yoruba people, where the plant is locally known as *"Egbo ifon"* [19]. Traditional healers in South-Western

TABLE 1.1
Traditional Uses of Some Species of the Genus *Olax*

Plants	Part(s) Used	Traditional Uses	Local Name(s)	Region	References
Olax mannii	Leaves	Cancer and inflammatory conditions	Tsadar birii (in Northern part of Nigeria)	Nigeria,	[60]
	Leaves and roots	Fever, yellow fever and snakebite	*Ngborogwu arua* in Ibo and *Tsada birii* in Hausa	Nigeria	[12]
	Root barks	Convulsions			
	Roots	Depression	*Tsada biri* in Hausa	Tribes of Kaduna State, Nigeria	[14]
Olax subscorpioidea	Roots	Irregular menstrual period	*Eza-bor* in Ogba/Egbema/ Ndoni Ethnics	Rivers State Nigeria	[16]
	Whole plant	Mental illnesses	*Gwano kurmi*	Sabo wuse in Niger state, Nigeria	[22]
	Roots	Asthma	*Ifon*	south western Nigeria	[23]
	Leaves, stem and roots	Jaundice and yellow fever	*Uburubu* in Ogoni people	River state Nigeria	[12]
	Roots	Convulsion			
	Stem	Toothache			
	Roots, leaves, stem bark and twigs	Yellow fever, jaundice, guinea worm, toothache, venereal diseases and mental disorders	*Ifan, Ifon* in Yoruba, Nigeria.	Nigeria	[9]
	Stem/root	Daily cleaning of the teeth, dental healthcare	*Egbo ifon*	Mushin Lagos state, Nigeria	[19]
	Whole plant	Constipation, yellow fever, jaundice, venereal diseases, Guinea worm and spice	/	Cameroon	[61]

(Continued)

TABLE 1.1 (Continued)
Traditional Uses of Some Species of the Genus Olax

Plants	Part(s) Used	Traditional Uses	Local Name(s)	Region	References
	Roots	Anaemia and intestinal worms	/	Northern and South-Eastern of Ivory cost	[62]
	Leaves, stem and roots	Jaundice and yellow fever	Uburubu (Ogoni)	Indigenes of Rivers State of Nigeria	[18]
	Roots	Convulsion Toothache			
	Roots	Malaria	/	Southern Benin	[63]
	Leaves	Malaria	/	Moronou (Ivory cost)	[64]
	Bark	Spices, prevents childhood diseases	Olom bekoé	Eton, Cameroon	[65]
	Roots, stems, branches and leaves	Dermatosis, fever, jaundice, rheumatism, colic, blennorrhagia, syphilis, arthritis and mental illness,	/	/	[3]
Olax viridis	Bark and roots	Ulcers, treatment of venereal disease, craw - craw and ring worm	/	West Africa	[26]
	Roots	Sleeping sickness, as a febrifuge, anti-diarrhea and in treatment of febrile headache	/	Northern of Nigeria	

(Continued)

TABLE 1.1 (Continued)
Traditional Uses of Some Species of the Genus Olax

Plants	Part(s) Used	Traditional Uses	Local Name(s)	Region	References
Olax scandens Roxb	Leaves	Headache	/	Western Ghats (Southern India)	[31]
	Leaves	Fever, constipation, cough, headache and mouth ulcer	Badru	India	[10]
	Stem bark	Fever and cough	Kaattu pavalam	Tamil Nadu (India)	[6]
Olax imbricata	Roots	Diabetes and cancer	/	Phu Yen province (south-central of Vietnam)	[32]
	Bark	Anemia and diabetes	/	Kuravas and Irulas (India)	[33]
Olax gambecola	Whole plant	Malaria	/	Bete Populations of Issia (Ivory cost)	[65]
	Roots	Oedema of pregnancy and constipation	Ekpenyong	Cross River State, Nigeria	[28]
	Whole plant	Ovarian cyst	/	Upper Nyong valley, Cameroon	[29]
	Roots	Aphrodisiac	/	Nimba mounts, Liberia	[30]

FIGURE 1.1 *Olax subscorpioidea* (West Region of Cameroon, Noun division).

Nigeria use *O. subscorpioidea* to treat diabetes mellitus [20]. Sickle cell disease is treated using the whole plant *O. subscorpioidea* in Southern Nigeria, where it is locally known under the following names: "*ifon*", "*igbula*", "*gwanorafi*" and "*gwanonraafi*" [21]. Inhabitants of Sabo Wuse heal mental illness using *O. subscorpioidea* (whole plant), locally called "*Gwano kurmi*" [22]. Moreover, roots of *O. subscorpioidea* are used in asthmatic conditions in South West Nigeria [23]. In Ghana, the stems are also used as a chewing stick to alleviate toothache [24]. In the Ivory Coast traditional medicine, this plant is used in the treatment of many diseases including jaundice and hepatitis [25]. The mixture of *O. subscorpioidea* with palm wine is locally called "*bandji*" for their traditional therapeutic utilization [25].

1.2.3 *OLAX VIRIDIS*

Olax viridis Oliv. is a shrub commonly found in the tropics. It is widely used for healthcare purposes. In West Africa, the pulverized bark and roots are used as a dressing for ulcers and treatment of venereal diseases, craw-craw and ring worm. In the northern part of Nigeria, the roots are used in the treatment of sleeping sickness, as a febrifuge and antidiarrheal and in the treatment of febrile headache. The leaves are used as a remedy for cough, fever and wound healing. A decoction of the leaves and twigs is also used as mouthwash and for toothache. The oil is usually placed in the tooth hollow prior to extraction. The crushed bark is applied to sores on domestic animals and used in the form of palm wine infusion as a remedy for typhoid fever [26]. *O. viridis* is used by Nsukka people in Nigeria as a hepatoprotective plant, and it is called "*Atu ndi umoha*" in this community [27].

1.2.4 *OLAX GAMBECOLA*

Commonly known as "*Ekpenyong*", *Olax gambecola* has been used widely by the traditional healers of Cross River State, Nigeria who found it useful in reducing fat (indicated by swollen ankles) during pregnancy. Preparation of the medicine involves boiling the roots of *O. gambecola* in water for at least 1 h with a two to three cupful of the water extract, and this was drunk daily. The roots have also been used by traditional healers in the treatment of constipation [28]. Local people of upper Nyong valley forest in Cameroon use the whole plant to cure ovarian cysts [29]. The maceration of its washed and cut roots for about 24 h in water or beer is used as aphrodisiac at Nimba Mount in Liberia [30]. In Ghana, its stems are used as a chewing stick to alleviate toothache [21].

1.2.5   OLAX SCANDENS

Olax scandens Roxb. is a shrub distributed throughout tropical India; it is used to cure fever, constipation, cough, headache and mouth ulcer [10]. Boiled leaves are tied onto the forehead one to two times to get relief from headache [31]. Known as "*Kaattu pavalam*", the decoction of *O. scandens* stem bark is taken internally to treat fever and cough by people of Tamil Nadu, India [6].

1.2.6   OLAX IMBRICATA

Olax imbricata is widely distributed in Phu Yen province, south-central Vietnam and has been used locally as a traditional remedy for diabetes and cancer [32]. Information collected from tribal groups called the *Kuravas* and *Irulas* (in India) about the traditional use of certain plants led to the observation that the bark of *O. imbricata* is used to cure anemia and diabetes in their locality [33].

1.3   PHYTOCHEMISTRY OF PLANTS OF THE GENUS *OLAX*

Many reports highlight the chemical composition of some plants of the genus *Olax* through phytochemical screenings. Thorough examination of these reports revealed the presence of several classes of compounds such as saponins, tannins, flavonoids, phenolic compounds, alkaloids and steroids [10,34]. To date, a few reports on the chemical characterization demonstrate that saponins and flavonoids are the main classes of compounds found in this genus. While a total of 46 compounds have been isolated and identified from the genus *Olax* 1981–2019, it should be noted that most isolated compounds were from two species, i.e., *O. mannii* and *O. imbricata*. The structures of these secondary metabolites were established by the use of NMR spectroscopic methods, followed by comparison of their data to those found in the literature; however, some of them required more analytical techniques in order to confirm their structures; especially electrospray ionization mass spectrometry (ESI-MS) was used for the purpose [32]. The structures of the isolated compounds are presented in Figures 1.1–1.5.

1.3.1   SAPONINS OF THE GENUS *OLAX*

This class of secondary metabolites is the most prevalent in the *Olax genus*; a total of 12 (Figure 1.2) have been found in *O. andronensis*, *O. glabriflora* and *O. psittacorum* [35]; *O. obtusifolia* [36] and *O. imbricata* [37]. From a methanol extract of the leaves, roots and barks of *O. andronensis*, *O. glabriflora* and *O. psittacorum*, a new oleanane type saponin namely olaxoside (**1**) was isolated by Forgacs

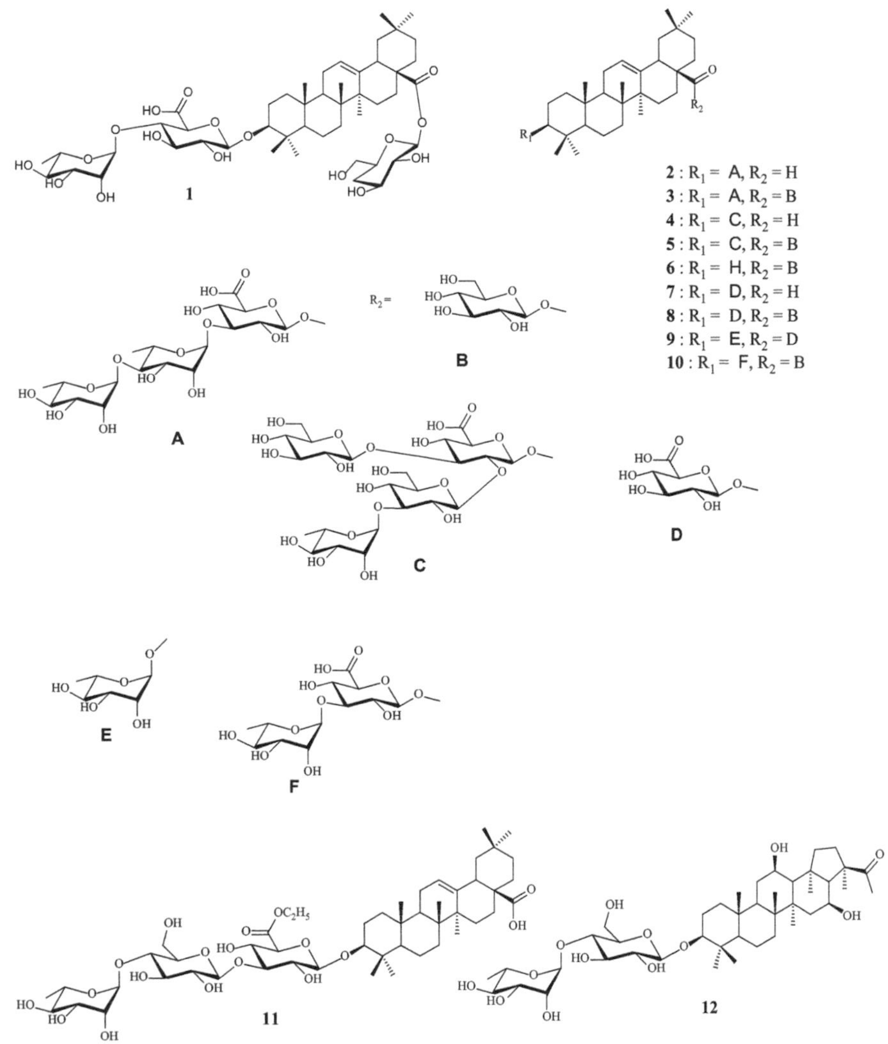

FIGURE 1.2 Saponins from the *Olax* genus.

and Provost (1981) [35]. The chemical investigation of the root bark methanol-
H$_2$O (80:20) extract of *O. obtusifolia* De Wild. was undertaken by Pertuit et al.
(2018) leading to the isolation of four new triterpenoid saponins by means of
repeated medium pressure liquid chromatography (MPLC) on normal and RP-18
silica gel and semi-preparative HPLC using RP-18 silica gel. These compounds
were elucidated as 3-*O*-α-L-rhamnopyranosyl-(1→4)-α-L-rhamnopyranosyl-
(1→3)-β-D-glucuronopyranosyl oleanolic acid (**2**), 3-*O*-α-L-rhamnopyranosyl-
(1→4)-α-L-rhamnopyranosyl-(1→3)-β-D-glucuronopyranosyloleanolic
acid 28-*O*-β-D-glucopyranosyl ester (**3**), 3-*O*-α-L-rhamnopyranosyl-(1→3)-β-D-
glucopyranosyl-(1→2)-[β-D-glucopyranosyl-(1→3)]-β-D-glucuronopyranosyloleanolic
acid (**4**) and 3-*O*-α-L-rhamnopyranosyl-(1→3)-β-D-glucopyranosyl-(1→2)-[β-D-

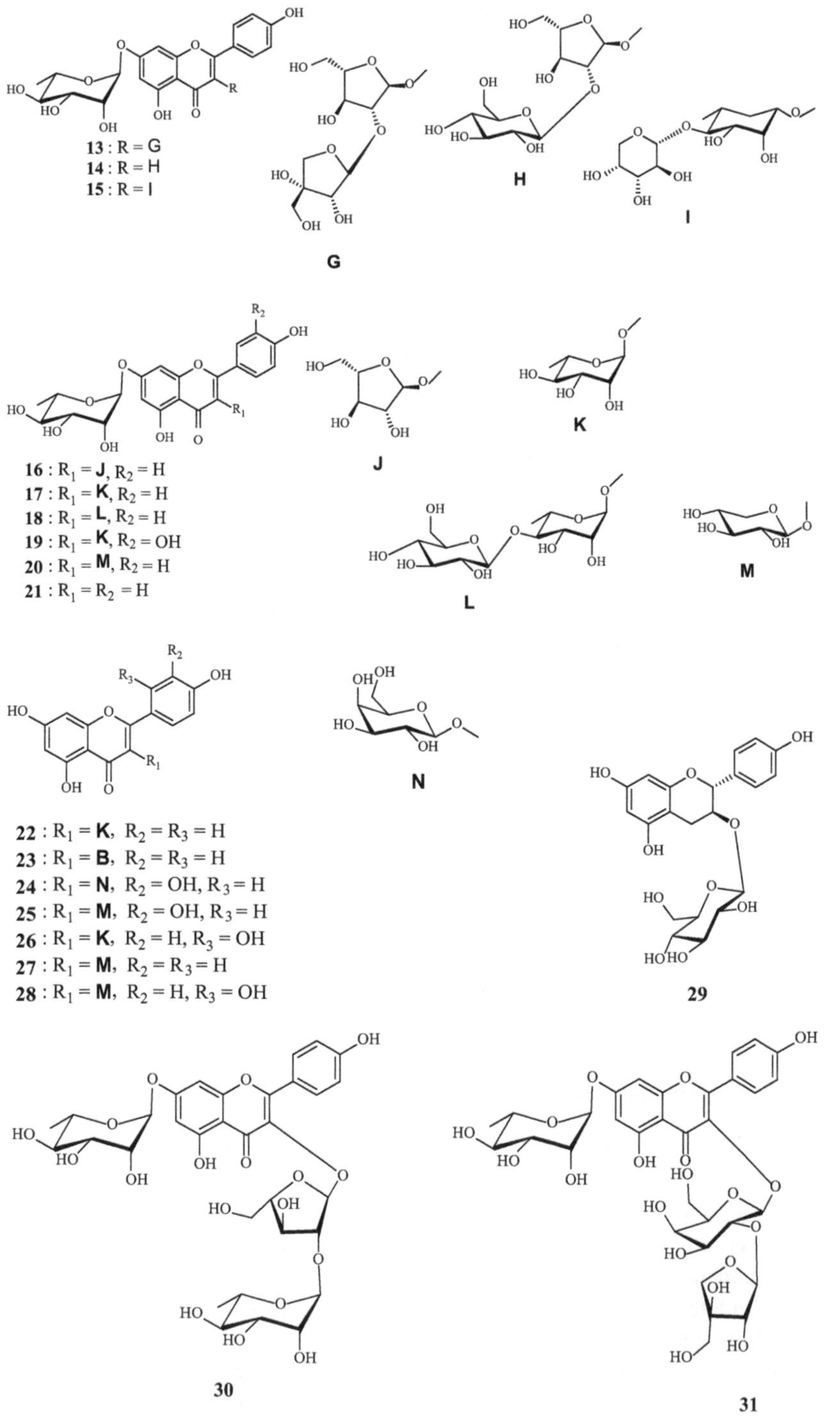

FIGURE 1.3 Flavonoids from the *Olax* genus.

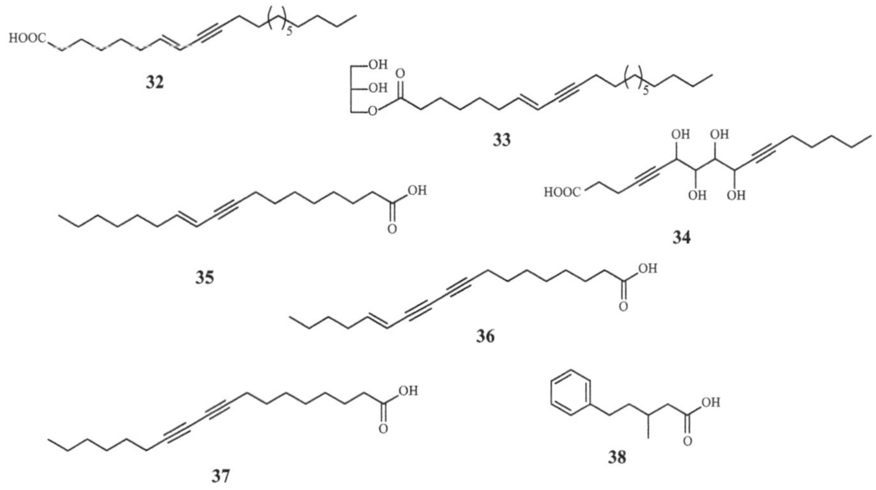

FIGURE 1.4 Fatty acids from the *Olax* genus.

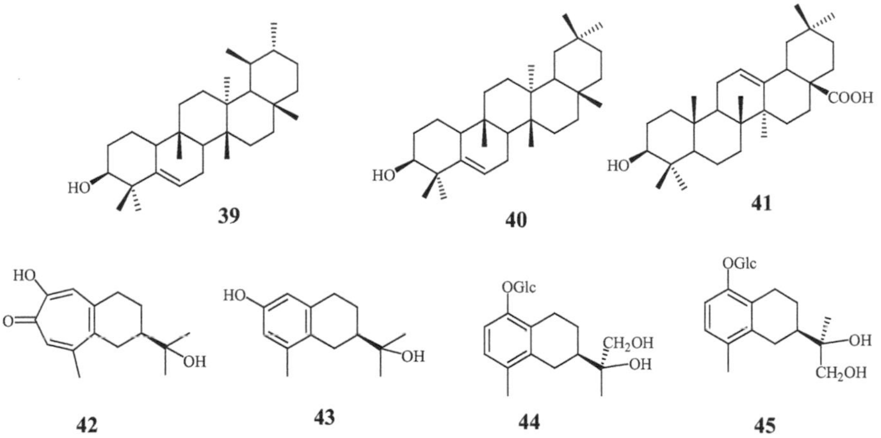

FIGURE 1.5 Terpenoids from the *Olax* genus.

glucopyranosyl-(1→3)]-β-D-glucuronopyranosyloleanolic acid 28-*O*-β-D-glucopyra nosyl ester (5) and five known derivatives, namely, oleanolic acid-28-*O*-β-D-glucopyr anosyl ester (**6**), 3-*O*-β-D-glucuronopyranosyloleanolic acid (**7**), 3-*O*-β-D-glucur onopyranosyloleanolic acid 28-*O*-β-D-glucopyranosyl ester (**8**), 3-*O*-α-L-rhamnopyranosyl-(1→3)-β-D-glucuronopyranosyloleanolic acid (**9**) and 3-*O*-α-L-rhamnopyranosyl-(1→3)-β-D-glucuronopyranosyloleanolic acid 28-*O*-β-D-glucopy ranosyl ester (**10**) [36]. Recently, Nga et al. (2019) used repeated column chroma-tography over silica gel of the methanol extract of *O. imbricata* roots to isolate three triterpenoid saponins: two oleanane 3-*O*-α-L-rhamnopyranosyl-(1→4)-β-D-glucopyranosyl-(1→3)-6′- *O*-ethyl-β-D-glucuronyl oleanolic acid (**11**) and oleanolic acid 28-*O*-β -D-glucopyranoside (**6**) and a new modified hopane skeleton that was given a specific name spergulacin (**12**) [37].

1.3.2   FLAVONOIDS OF THE GENUS *OLAX*

The *n*-butanol fraction of a methanol leaf extract of *Olax mannii* was subjected to vacuum-liquid chromatography (silica gel, sintered funnel), Sephadex LH-20 column and semi-preparative HPLC by Okoye et al. (2015) and yielded 17 flavonoids [38], three new derivatives named olamannoside A (**13**), olamannoside B (**14**) and olamannoside C (**15**) and 14 known ones named kaempferol 3-*O*-α-L-arabinofuranosyl-7-*O*-α-L-rhamnopyranoside (**16**), kaempferol 3,7-*O*-α-L-dirhamnopyranoside (kaempferitin) (**17**), kaempferol 3-*O*-[β-D-glucopyranosyl-(1–4)-α-L-rhamnopyranoside]-7-*O*-α-L-rhamnopyranoside (**18**), quercetin 3,7-O-α-L-dirhamnopyranoside (**19**), kaempferol-3-*O*-β-D-xylopyranosyl-7-*O*-α-L-rhamnopyranoside (**20**), kaempferol 7-*O*-α-L-rham nopyranoside (**21**), kaempferol 3-*O*-α-L-rhamnopyranoside (**22**), kaempferol 3-*O*-β-D-glucopyranoside (**23**), quercetin-3-*O*-β-D-galactopyranoside (**24**), quercetin 3-*O*-β-D-xylopyranoside (**25**), morin 3-*O*-α-L-rhamnopyranoside (**26**), kaempferol 3-*O*-β-D-xylopyranoside (**27**), Morin 3-*O*-β-D-xylopyranoside (**28**) and (+)-afzel-echin 3-*O*-β-D-glucopyranoside (**29**) [38]. Further studies carried out by Okoye et al. (2016) led to the isolation of further new derivatives named olamannoside D (**30**) and olamannoside E (**31**) [39]. The structures are presented in Figure 1.3.

1.3.3   FATTY ACIDS OF THE *OLAX* GENUS

Some unsaturated fatty acids were isolated from three species of the genus *Olax* namely *O. imbricata* [40], *O. dissitiflora* [41] and *O. mannii* [42] (Figure 1.3). Silica gel column chromatography of the ethyl acetate extract of *O. imbricata* root extract carried out by Nga et al. (2018) led to the isolation of three novel fatty acids identified as (*E*)-henicos-7-en-9-ynoic acid (**32**), (*E*)-henicos-7-en-9-ynoylglycerol (or 2,3-dihydroxypropyl (*E*)- henicos-7-en-9-ynoate) (**33**) and 6,7,8,9-tetrahydroxyhexa-deca-4,10-diynoic acid (**34**) [40]. From the dichloromethane extract of *O. dissitiflora*, three known fatty acids were partially isolated through the use of vacuum liquid chromatography, i.e., *trans*-11-octadecen-9-ynoic acid, also known as santalbic acid or ximeninic acid (**35**) and a mixture of exocarpic acid (**36**) and octadec-9,11-diynoic acid (**37**) [41]. Furthermore, chemical study of the petroleum ether extract of the leaves of *O. mannii* yielded the isolation and identification of one novel fatty acid (*E*)-3- methyl-5-phenyl-2-pentenoic acid (**38**) [42] (Figure 1.4).

1.3.4   TERPENOIDS OF THE GENUS *OLAX*

Compounds **39** and **40** representing oleanane-type and ursane-type triterpenoids, respectively, were reported for the first time from the species *Olax mannii* by Sule et al. (2011) [13]. One sesquiterpenoid and three 1,2,3,4-tetrahydronaphthalene derivatives (**42–45**) were identified from *O. imbricata* root extract by Nguyen [32]. A chemical investigation of an acetone extract of *O. mannii* using column chromatography, preparative TLC and recrystallization processes afforded a new derivative called rhoiptelenol (**39**) and glutinol (**40**) [13]. Oleanolic acid (**41**) was isolated from root bark of *O. obtusifolia* methanol-H_2O extract (80:20) [36]. Phytochemical investigations by means of repeated silica gel column chromatography on the methanol

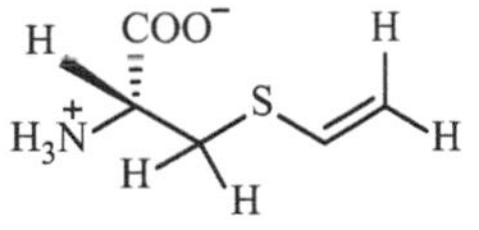

46

FIGURE 1.6 Amino acid from the *Olax* genus.

extract of *O. imbricata* roots yielded four previously unreported metabolites identified as olaximbriside A (**42**), olaximbriside B (**43**), olaximbriside C (**44**) and olaximbriside D (**45**) [32] (Figure 1.5).

1.3.5 AMINO ACID OF THE *OLAX* GENUS

The application of a typical preparative extraction procedure by Peter et al. (1993) resulted in the isolation and identification of an amino acid derivative, *S*-ethenyl cysteine (**46**) from the leaves of *O. phyllanthi* [43] (Figure 1.6).

1.4 BIOLOGICAL INVESTIGATIONS

Plants continue to be used worldwide to cure diseases, but their extracts need to undergo biomedical scientists' search to understand the complex aspect of traditional healing [44].

1.4.1 BIOLOGICAL ACTIVITIES OF *OLAX* PLANT SPECIES EXTRACTS

Exploitation of useful information collected from traditional healers motivated researchers to focus on biological activities of plants belonging to the genus *Olax*. These studies include antibacterial, antifungal, anti-inflammatory, antioxidant, DPPH free radical scavenging, antinociceptive and antidepressant-like effect activities [5,10,15,26].

1.4.1.1 Antimicrobial Activity of *Olax* Plant Species Extracts

The root extracts of *Olax scandens* have been investigated recently to assess their antimicrobial activity. Chloroform, methanol and aqueous extracts displayed significant activities against *Bacillus subtilis* and *S. aureus*. Among the fungal strains, only *Aspergillus niger* and *Saccharomyces cerevisiae* were affected by these extracts [10]. The antimicrobial activity carried out on the leaf extract of *O. subscorpioidea* led to a moderate sensitivity against *Salmonella typhi* and *Escherichia coli*, and interesting activities were obtained from methanol extract against *S. aureus* and *Klebsiella pneumoniae* [45]. Investigation on ethanolic and aqueous stem extracts of this plant conducted some years ago demonstrated that the ethanolic extract was active against many strains tested, while the aqueous extract was only active against two bacterial strains *Pseudomonas sp.* and *Proteus vulgaris* [46]. It has been reported recently that root extracts of *O. subscorpioidea* possess good antibacterial activity against *E. coli*, *Proteus* and *Shigella* [47]. The stem and fruit aqueous extracts of *O. psittacorum* were tested against six bacterial strains including *S. aureus*, *Bacillus stearothermophilus*,

Pseudomonas aeruginosa, Vibrio cholera, E. coli and *Acinetobacter baumannii*; only two out of the six underwent the effects of fruit aqueous extracts, whereas stem aqueous extract exhibited no antibacterial activity [5].

The methanol-dichloromethane extract of *O. subscorpioidea* fruits was assessed by Dzoyem et al. (2014) for its antifungal activity against *Candida albicans, C. parapsilosis, C. tropicalis, C. krusei, C. lusitaniae, Cryptococcus neoformans* and two clinical isolates namely *C. guilliermondii* and *C. glabrata*. The assay was investigated following the rapid 2-(4-iodophenyl)-3-(4-nitrophenyl)-5-phenyltetrazolium chloride (INT) colorimetric assay using nystatin as a positive control. This extract exhibited a significant *in vitro* activity against *C. albicans* and *C. tropicalis* with MIC values of 0.09 and 0.05 mg/mL, respectively. The *in vivo* test demonstrated a decrease in fungal load in rats' kidneys after administration of the plant extract [15].

1.4.1.2   Anti-Inflammatory Activity of *Olax* Plant Species Extracts

After administration of different concentrations of hexane root extract of *O. viridis*, a moderate inhibition of paw edema induced by egg albumen was obtained compared to the effect of acetyl salicylic acid. These results were promising in the way that the ulcerogenic activity was lower than the one of ASA 1.83 and 2.20 mg/kg, respectively [26]. Similarly, evaluation of ethanol and aqueous extracts of *O. subscorpioidea* led to a significant reduction of the paw size resulting from the induced adjuvant-induced arthritis in rats [48]. Another way for inflammation to occur in an organism is related to the lysosomal enzymes' release; besides, ethanol leaf extract of *O. subscorpioidea* has been tested for this purpose through the bovine red blood cell as physiological condition. It showed a percentage of inhibition of 87.5% of lysosomal membrane, an activity comparable to that of indomethacin [49].

1.4.1.3   Antivenom Activity of *Olax* Plant Species Extracts

The potential of *O. viridis* to be effective on a snake bite has been investigated by studying some biological parameters such as pulse rate and blood pressure as well as antipyretic activity in rats. As a result, the methanol extract exhibited a significant activity [50].

1.4.1.4   DPPH Free Radical Scavenging Activity
of *Olax* Plant Species Extracts

Olax psittacorum leaf extracts (methanol, ethanol and hexane) were assessed for their capacity to scavenge free radicals. For this purpose, DPPH was used as the radical and ascorbic acid as the positive control. The percentages of inhibition were 91% and 96%; 97% and 96%; 77% and 96% for methanol, ethanol and hexane extracts, respectively. From their IC_{50} values, the ethanol extract was the most active (IC_{50} 27.56 µg/mL) [51].

1.4.1.5   Antioxidant Activity of *Olax* Plant Species Extracts

The antioxidant activity of *Olax nana* hydromethanolic (1:1) leaf extract assessed through DPPH, ABTS and H_2O_2 led to significant results compared to those of the positive control (ascorbic acid). IC_{50} values were 71.46 and 31.31 µg/mL, 72.55 and 32.37 µg/mL and 92.33 and 22.01 µg/mL for each method, respectively, relative to that of ascorbic acid for each one [52]. A methanol extract of *Olax zeylanca* leaves led to a moderate antioxidant capacity expressed as ascorbic acid equivalent per gram (AAE/g) of dried leaves [53]. Aqueous extracts of *Olax subscorpioidea leaves* were assessed for this activity through DPPH, ABTS (calculated as trolox equivalent antioxidant capacity), ion chelation and lipid peroxidation and thiobarbituric acid reaction assays. From these IC_{50} values, this extract showed a significant antioxidant activity (19.74 mg AEE/g, 0.68 mmol TEAC/100 g, 255.84 mg/mL and 111.03 g/mL, respectively, for the methods above) [54]. Recently, similar results were obtained from the aqueous extract of stem and fruits of *Olax psittacorum* [4]. Another assay performed on essential oil of *Olax acuminata* leaves showed a significant antioxidant activity through DPPH and superoxide assays [34].

The antioxidant activity of the leaf extracts of *O. scandens* was assessed through DPPH scavenging activity, superoxide activity and iron chelating potential by Pranaya and Akila (2019). In DPPH scavenging, petroleum ether, ethyl acetate and methanolic extracts displayed IC_{50} values of 990, 504 and 226 µg/mL, respectively, against 66 µg/mL for ascorbate. For the superoxide radical evaluation, these extracts exhibited IC_{50} values of 1070, 495 and 185 µg/mL, respectively, and relatively to 60 µg/mL for quercetin, and in iron-binding potential, IC_{50} values were 980, 485 and 287 µg/mL, respectively, compared to that of ethylenediamine tetracetate at 65 µg/mL [55].

1.4.1.6 Hypoglycemic Activity of *Olax* Plant Species Extracts

The plasma glucose was decreased after 6 weeks when streptozotocin-induced diabetic rats were injected with different concentrations of *O. mannii* aqueous root extracts. Besides, these rats received a concentration of 1.5 mg/kg of glibenclamide and different concentrations of aqueous root extract of the plant (500, 1000 and 1500 mg/kg). After 6 weeks, the plasma glucose level decreased from 180 to 78 mg/dL and 178 to 79 mg/dL for glibenclamide and plant extract (1500 mg/kg), respectively [12]. The α-glycosidase activity, which is also associated with a rise of plasma glucose, has been affected by the methanol leaf extract of *O. nana*, with dissociation constants of 31.11 and 51.01 µg/min of plant extract and that of positive control (acarbose), respectively [52].

1.4.1.7 Hypolipidemic Activity of *Olax* Plant Species Extracts

In order to validate the hypolipidemic effect of *O. subscorpioidea*, hyperlipidemia was induced by hypercholesterolemia in rats, and they were treated using atorvastatin as a positive control and root ethanol (50%) extract of *O. subscorpioidea*. After 2 weeks, significant results were obtained by measuring some lipid parameters in rats (total cholesterol, triglyceride, high density lipoproteins and low-density lipoproteins). For total cholesterol, the initial value in induced rat was 122.20 mg/dL, and resulting values in treated rats were 81.20, 99.76 and 112.92 mg/dL, respectively, for atorvastatin 10 mg, *O. subscorpioidea* ethanol extract 200 mg and 400 mg [56].

1.4.1.8 Acetylcholinesterase and Butyrylcholinesterase Inhibitory Activity of *Olax* Plant Species Extracts

A strong inhibitory effect was obtained from *O. nana* leaf extract (MeOH 70%), in fact, for this extract, a dissociation value of acetylcholinesterase of 20.94 µg against 16.43 µg of positive control (galantamine). A similar effect was shown with butyrylcholinesterase, making this extract a possible good candidate to fight against

Alzheimer's disease [53]. Investigations on *O. subscorpioidea* extract led to similar results (110.35 and 108.44 mg/mL for AChE and BChE IC_{50} values, respectively), and the inhibition was shown to be dose-dependent [54].

1.4.1.9 Other Activities of *Olax* Plant Species Extracts

Olax subscorpioidea leaf extract was evaluated for its ability to induce the mitochondrial membrane permeability transition pore, which was found to be successful. This experiment was investigated using spermine as a positive control, and the results were expressed in terms of change in absorbance of test media. Also, induction of mitochondrial membrane lipid peroxidation and an increase of mitochondrial phosphate release were observed [57]. Hydroethanol leaf extract of this plant was also found to possess antidepressant-like effect on mice by Adeoluwa et al. In fact, at doses of 6.25 and 12.5 mg/kg, there was a significant reduction of immobility in mice in a forced swimming test compared to the positive control (yohimbine) [58]. Similar activity on *n*-butanol fraction of leaf extract also showed antidepressant-like effect [59].

1.4.2 Biological Activities of Isolated Compounds

Some secondary metabolites isolated from plants of the genus *Olax* have been assessed for their biological activities. The flavonoids isolated from O. mannii were investigated, and only the compound 22 led to a promising antiproliferative effect [38]; olamnanosides D (**30**) and E (**31**) showed weak inhibition toward some microorganisms [39]. During the *in vitro* evaluation for potential cytotoxicity against three cancer cell lines using the modified MTT method, olaximbriside A (**42**) exhibited good activity against breast cancer MCF-7, HepG2 and LU cancer cell lines (liver cancer) with IC_{50} values of 16.5, 34.3 and 8.0 μM, respectively [32]. Kaempferol 3-*O*-α-L-rhamnopyranoside (**22**) was screened for cytotoxic activity against human K562 leukemia cells using quantitative IncuCyte video microscopy. As a result, it exhibited a significant antiproliferative effect at 50 μM. The same compound also produced a significant inhibition of K562 cellular metabolism at a concentration of 50 μM after 72 h of treatment. The cytotoxicity of the same compound on K562 cells was also assessed by trypan blue test, and the results obtained showed that it was nontoxic to K562 cells at 10 and 50 μM after 8 h of exposure but gave a significant toxicity after 24 h and that it was found to be dose-dependent with an IC_{50} of 15.2 μM. Furthermore, NF-κB assays were performed on the same compound, and it displayed a significant dose-dependent inhibition of NF-κB with an IC50 of 4.1 μM [38].

Larvicidal activity was conducted on three fatty acids, i.e., santalbic acid (**35**), exocarpic acid (**36**) and octadec-9,11-diynoic acid (**37**), isolated from *O. dissitiflora*. The results showed that santalbic acid possessed the lowest activity in comparison with the mixture of exocarpic acid and octadec-9,11-diynoic acid with an EC_{50} value 62.17 and 17.31 μg/mL, respectively [41]. Oleanolic acid 28-*O*-β-D-glucopyranoside (**6**) was evaluated for its potential α-glucosidase inhibitory effects using the Hakamata's method with some modification using acarbose as the positive control. This compound exhibited an activity with IC_{50} value 56.15 mM, while that of the positive control was 290.59 mM [37].

1.5 CONCLUSION

Many of the plants belonging to the genus *Olax* are used in traditional medicine especially in Cameroon, Nigeria, Ghana and India to cure many diseases. The extracts and pure compounds derived from some of these plants have exhibited interesting antimicrobial and cytotoxic properties, thus supporting their uses in traditional medicine. Phytochemical investigations have been carried out on only five species of this genus leading to the isolation and identification of 46 compounds including 18 previously unreported metabolites. Given that about half of the secondary metabolites isolated from plants of this genus are new, additional phytochemical and pharmacological studies are required to discover new and promising bioactive secondary metabolites.

Although all the species of the genus *Olax* discussed in this review are present in Cameroon, there are no documents dealing with the uses of some of these plants in traditional medicine in this country. Furthermore, the phytochemical investigation of a plant belonging to this genus and collected in Cameroon has not yet been achieved. This means that ethnobotanists, biologists and phytochemists in Cameroon must work urgently on plants of the genus *Olax* because they represent a potential source for the discovery of bioactive substances that can serve as hits and leads for new drugs.

REFERENCES

1. Assob, J.C.N., Kamga, H.L.F., Nsagha, D.S., Njunda, A.L., Nde, P.F., Songalem, E.A., et al. 2011. Antimicrobial and toxicological activities of five medicinal plant species from Cameroon traditional medicine. *BMC Complementary and Alternative Medicine* 70: 1–11.
2. World Health Organization. 2019. List of bacteria for which new antibiotics are urgently needed. Available at https://www.who.int/news-room/detail/27-02-2017-who-publishes-list-of-bacteria-for-which-new-antibiotics-are-urgently-needed. accessed June 25, 2019.
3. Konan K., N'guessan, J.D., Méité, S., Yapi, A., Yapi, H.F. and Djaman, A.J., 2013. *In vitro* antioxidant activity and phenolic contents of the leaves of *Olax subscorpioideae* and *Distemonanthus benthamianus*. *Research Journal of Pharmaceutical, Biological and Chemical Sciences* 4(4): 14–19.
4. Guantai, E.M. and Chibale, K., 2012. Natural product-based drug discovery in Africa: the need for intergration into modern drug discovery paradigms. In: *Drug discovery in Africa*. Germany: Springer, 101–126.
5. Majumder, R., Dhara, M., Adhikari, L., Ghosh, G. and Pattnaik, S., 2019. Evaluation of *in vitro* antibacterial and antioxidant activity of aqueous extracts of *Olax psittacorum*. *Indian Journal of Pharmaceutical Sciences* 81: 99–109.
6. Duraipandiyan, V., Ayyanar, M. and Ignacimuthu, S., 2006. Antimicrobial activity of some ethnomedicinal plants used by Paliyar tribe from Tamil Nadu, India. *BMC Complementary and Alternative Medicine* 6(35): 1–7.
7. Mahomoodally, M.F., 2013. Traditional medicines in Africa: An appraisal of ten potent African medicinal plants. *Evidence-Based Complementary and Alternative Medicine* 2013: 1–14.
8. Breteler, J.F., Baas, P., Boesewinkel, F.D., Bouman, F. and Lobreau-Callen, D., 1996. *Engomegoma* Breteler (Olacaceae) a new monotypic genus from Gabon. *Botanische Jahrbücher für Systematik, Pfl anzengeschichte und Pfl anzengeographie* 118: 113–132.

9. Olowokudejo, J.D., Kadiri, A.B. and Travih, V.A., 2008. An ethnobotanical survey of herbal markets and medicinal plants in Lagos states of Nigeria. *Ethnobotanical Leaflets* 12: 851–865.

10. Owk, A.K. and Lagudu, M.N., 2016. Evaluation of antimicrobial activity and phytochemicals in *Olax scandens* roxb. Roots. *An International Journal of Pharmaceutical Sciences* 7(2): 232–239.

11. Beche, L., Descloux, S. and Perrot, F., 2011. Projet hydroélectrique de Nachtigal, Etude biodiversité. *Inventaires flore et faune complémentaires* 67: 1–21.

12. Nwauche, K.T., Nwosu, G.U. and Ighorodje-Monago, C.C., 2017. Hypoglycemic activity of the aqueous root-back extract of *Olax mannii* on diabetic induced albino rats. *International Journal of Innovative Research and Advanced Studies* 4(6): 570–574.

13. Sule, M.I., Hassan, H.S., Pateh, U.U. and Ambi, A.A,. 2011. Triterpenoids from the leaves of *Olax mannii* Oliv. *Nigerian Journal of Basic and Applied Science* 19(2): 193–196.

14. Shehu, A., Magaji, M.G., Jamilu Yau, J. and Ahmed, A., 2017. Ethnobotanical survey of medicinal plants used for the management of depression by Hausa tribes of Kaduna State, Nigeria. *Journal of Medicinal Plants Research* 11(36): 562–567.

15. Dzoyem, J.P., Tchuenguem, R.T., Kuiate, J.R., Teke, G.N., Kechia, F.A. and Kuete, V., 2014. *In vitro* an *in vivo* antifungal activities of selected Cameroonian dietary spices. *BMC Complementary and Alternative Medicine* 14: 1–8.

16. Oladele, A.T. and Elem, J.A., 2018. Medicinal plants used to treat sexual diseases by Ogba/Egbema/Ndoni ethnics of Rivers State, Nigeria. *World News of Natural Sciences* 17: 16–38.

17. Soladoye, M.O., Amusa, N.A., Raji-Esan, S.O., Emmanuel C. Chukwuma, E.C. and Taiwo, A.A., 2010. Ethnobotanical survey of anti-cancer plants in Ogun State, Nigeria. *Annals of Biological Research* 1(4): 261–273.

18. Ajibesin, K.K., Bala, D.N. and Umoh, U.F., 2012. Ethnomedicinal survey of plants used by the indigenes of Rivers State of Nigeria. *Pharmaceutical Biology* 50(9): 1123–1143.

19. Orabueze, I.C., Amudalat, A.A. and Usman, A.A., 2016. Antimicrobial value of *Olax subscorpioideae* and *Bridelia ferruginea* on micro-organism isolates of dental infection. *Journal of Pharmacognosy and Phytochemistry* 5(5): 398–406.

20. Soladoye, M.O., Chukwuma, E.C. and Owa, F.P., 2012. An 'Avalanche' of plant species for the traditional cure of diabetes mellitus in South-Western Nigeria. *Journal of Natural Product and Plant Resources* 2(1): 60–72.

21. Amujoyegbe, O.O., Idu, M., Agbedahunsi, J.M. and Erhabor, J.O., 2016. Ethnomedicinal survey of medicinal plants used in the management of sickle cell disorder in Southern Nigeria. *Journal of Ethnopharmacology* 185: 347–360.

22. Ibrahim, J.A., Muazzam, I., Jegede, I.A., Kunle, O.F. and Okogun, J.I., 2007. Ethnomedicinal plants and methods used by Gwandara tribe of Sabo wuse in Niger state, Nigeria, to treat mental illness. *African Journal of Traditional, Complementary and Alternative Medicines* 4(2): 211–218.

23. Sonibare, M.A. and Gbile, Z.O., 2008. Ethnobotanical survey of anti-asthmatic plants in south western Nigeria. *African Journal of Traditional, Complementary and Alternative Medicines* 5(4): 340–345.

24. Portères, R., 1974. Les baguettes végétales machées servant de frotte-dents (fin). *Journal d'agriculture traditionnelle et de botanique appliquée* 21: 111–150.

25. Osuntokun, O.T. and Omolola, A.Y., 2019. Efficacy of Nigerian medicinal plant (*Olax subscorpioideae* Oliv.) Root extract against surgical wound isolates. *American Journal of Microbiology and Biochemistry* 2(1): 001–011.

26. Ajali, U., and Okoye, F.B.C., 2009. Antimicrobial and anti-inflammatory activities of *Olax viridis* root bark extracts and fractions. *International Journal of Applied Research in Natural Products* 2(1): 27–32.

27. Nwaigwe, C.U., Madubunyi, I.I., Udem, S.C. and Nwaigwe, C.O., 2012. Methanolic root extract of *Olax viridis* protects the liver against acetaminophen-induced liver damage. *Research Journal of Medicinal Plant* 6(5): 395–405.

28. Parry, O., Okwuasaba, F.K., Ekpenyonga, K.I. and Ashrafa, C.M., 1986. Effects of *Olax gambecola* methanol extract on smooth muscle and rat blood pressure. *Journal of Ethnopharmacology* 18: 63–88.

29. Jiofack, T., Ayissi, I., Fokunang, C., Guedje, N. and Kemeuze, V., 2009. Ethnobotany and phytomedicine of the upper Nyong valley forest in Cameroon. *African Journal of Pharmacy and Pharmacology* 3(4): 144–150.

30. Adam J.G., 1971. Quelques utilisations de plantes par les Manon du Libéria (Monts Nimba). *Journal d'agriculture tropicale et de botanique appliquée* 18(9–10): 372–378.

31. Jeyaprakash, K., Ayyanar, M., Geetha, K.N. and Sekar, T., 2011. Traditional uses of medicinal plants among the tribal people in Theni District (Western Ghats), Southern India. *Asian Pacific Journal of Tropical Biomedicine* 1: S20–S25.

32. Nguyen, H.T.M., Vo, N.T., Huynh, S.T.M., Do, L.T.M., Tip-pyang, T.A.S., Phan, C.T.D., et al., 2019. A sesquiterpenoid tropolone and 1, 2, 3, 4-tetrahydronaphthalene derivatives from *Olax imbricata* roots. *Fitoterapia*, 132: 1–6.

33. Arulappan, M.T., Britto, S.J., Ruckmani, K. and Kumar, R.M., 2015. An ethnobotanical study of medicinal plants used by ethnic people in Gingee hills Villupuram District, Tamilnadu, India. *American Journal of Ethnomedicine* 2(2): 84–102.

34. Chetiaa, B. and Phukan, P., 2014. Chemical composition and antioxidant activities of the essential oil of *Olax acuminata*. *Journal of Essential Oil Bearing Plants*, 17(4): 696–701.

35. Forgacs, P. and Provost, J., 1981. Olaxoside, a saponin from *Olax andronensis*, *Olax glabriflora* and *Olax psittacorum*. *Phytochemistry* 20(7): 1689–1691.

36. Pertuit, D., Mitaine-Offer, A.C., Miyamoto, T., Tanaka, C., Delaude, C. and Lacaille-Dubois, M. A., 2018. Triterpene saponins of the root bark of *Olax obtusifolia* De Wild. *Phytochemistry Letters* 28: 174–178.

37. Nga, V.O, Suong, H.T.M., Huong, N.T.M. Huy, D.T. and Phung, N.K.P., 2019. Triterpenoid glycosides from *Olax imbricata*. *Science & Technology Development Journal* 22(3): 324–334.

38. Okoye F.B.C., Sawadogo, W.R., Sendker, J., Aly, A.H., Quandt, B., Wray, V., et al., 2015. Flavonoid glycosides from *Olax mannii*: Structure elucidation and effect on the nuclear factor kappa B pathway. *Journal of Ethnopharmacology* 176: 27–34.

39. Okoye, F.B.C., Ngwoke, K.G., Debbaba, A., Osadebe, P.O. and Proksch, P., 2016. Olamannosides D and E: Further kaempferol triglycosides from *Olax mannii* leaves. *Phytochemistry Letters* 16: 152–155.

40. Nga, V.T., Huong, N.T.M. and Phung, N.K.P., 2018. Fatty compounds from *Olax Imbricata*. *4th International Conference on Green Technology and Sustainable Development (GTSD)*, Ho Chi Minh City, Vietnam, 538–540.

41. Mavundza, E.J., Chukwujekwu, J.C.R. Maharaj, J.F., Finnie, F.R., Heerden, J.V. and Staden, V., 2016. Identification of compounds in *Olax dissitiflora* with larvacidal effect against *Anopheles arabiensis*. *South African Journal of Botany* 102: 1–3.

42. Sule M.I., Haruna, A.K. Pateh, U.U. Ahmadu, A.A. Ambi, A.A. and Sallau, M.S., 2005. Phytochemical investigation of leaf, fruit and root bark of *Olax mannii* Oliv. Olacaceae. *ChemClass Journal* 2: 22–24.

43. Peter, P.T., John, S.P., Edvin, R. and Emilio, L.H., 1993. *S*-ethenyl cysteine; an amino acid from *Olax phyllanthi*. *Phytochemistry* 34: 651–659.

44. Swapna, G., Shaheeda, S., Sai, S., Swapna, P. and Sirisha, V., 2016. Bioassay guided evaluation of herbal drugs and medicinal plants. *Journal of Medical Pharmaceutical and Allied Sciences* 93–98.

45. Abdulazeez, I., Lawal, A. Y. and Aliyu, S., 2015. Phytochemical screening and antimicrobial activity of the solvents' fractionated leaves extract of *Olax subscorpioideae*. *Journal of Chemical and Pharmaceutical Research* 7(9): 22–26.

46. Ayandele, A.A. and Adebiyi, A.O., 2007. The phytochemical analysis and antimicrobial screening of extracts of *Olax subscorpioideae*. *African Journal of Biotechnology* 6(7): 868–870.

47. Banjo O.A., Ashidi J.S., Oyelami L.A., Onajobi I.B and Onifade F.T., 2019. Phytochemical components and antibacterial activity of *Olax subscorpioidea* root on selected bacterial pathogens. *FUOYE Journal of Pure and Applied Sciences* 3: 355–364.

48. Ezeani, N.N., Ibiam, U.A., Orji, I.O., Alum, E.U., Aja, P.M., Igwenyi, I.O., et al., 2018. Effects of aqueous and ethanol root extracts of *Olax subscorpioideae* on inflammatory parameters in complete Freund's adjuvant-collagen type II induced arthritic albino rats. *Journal of Pharmaceutical Chemistry and Pharmacology* 2(1): 1–10.

49. Oyedapow, O.O., Akindele, V. R. and Okunfolami, O.K., 1997. Effects of extracts of *Olax subscorpioideae* and *Aspilia africana* on bovine red blood cell. *Phytotherapy Research* 11: 305–306.

50. Omale, J., Ebiloma, U.G. and Ogohi, D.A., 2013. Anti-venom studies on *Olax viridis* and *Syzygium guineense* extracts against Naja katiensis venom in rats. *American Journal of Pharmacology and Toxicology* 8(1): 1–8.

51. Sahu, R.K., Kar, M. and Routray, R., 2013. DPPH free radical scavenging activity of some leafy vegetables used by tribals of Odisha, India. *Journal of Medicinal Plants Studies* 1(4): 21–27.

52. Ovais, M., Muhammad, A., Khalil, A.T., Shah, S.A., Jan, M.S., Raza, A., et al., 2018. HPLC-DAD finger printing, antioxidant, cholinesterase, and α-glucosidase inhibitory potentials of a novel plant *Olax nana*. *BMC Complementary and Alternative Medicine* 18(1): 1–13.

53. Gunathilake, K.D.P.P. and Ranaweera, K.K.D.S., 2016. Antioxidative properties of 34 green leafy vegetables. *Journal of Functional Foods* 26: 176–186.

54. Jamiu, A.S. and Ayodeji, A.O., 2017. Aqueous extract of *Securidaca ongipendunculata* Oliv. and *Olax subscropioideae* inhibits key enzymes (acetylcholinesterase and butyrylcholinesterase) linked with Alzheimer's disease *in vitro*. *Pharmaceutical Biology* 55: 252–257.

55. Pranaya, P. and Akila, D.D., 2019. Study of *in-vitro* antioxidant activity of various extracts of aerial parts of *Olax scandens*. *International Journal of Research in Pharmaceutical Sciences* 10(4): 3475–3480.

56. Gbadamosi, I.T., Raji, L.A., Oyagbemi, A.A. and Omobowale, T.O., 2017. Hypolipidemic effects of *Olax subscorpioideae* Oliv. Root extract in experimental rat model. *African Journal of Biomedical Research* 20: 293–299.

57. Adegbite, O.S., Akinsanya, Y.I., Kukoyi, A.J., IyandaJoel, W.O., Daniel, O.O. and Adebayo, A.H., 2015. Induction of rat hepatic mitochondrial membrane permeability transition pore opening by leaf extract of *Olax subscorpioideae*. *Pharmacognosy Research* 7: S63–S68.

58. Adeoluwa, O.A., Aderibigbe, A.O. and Bakre, A.G., 2015. Evaluation of antidepressant-like effect of *Olax Subscorpioidea* Oliv. (Olacaceae) extract in mice. *Drug Research* 65: 306–311.

59. Adeoluwa, A.O., Aderibigbe, O.A., Agboola, I.O., Olonode, T.E. and Ben-Azu, B., 2019. Butanol fraction of *Olax Subscorpioidea* produces antidepressant effect: Evidence for the involvement of monoaminergic neurotransmission. *Drug Research* 69: 53–60.

60. Abubakar, M.S., Musa, A.M., Ahmed, A. and Hussaini, I.M., 2007. The perception and practice of traditional medicine in the treatment of cancers and inflammations by the Hausa and Fulani tribes of Northern Nigeria. *Journal of Ethnopharmacology* 111: 625–629.

61. Kuete, V. and Efferth, E., 2015. African flora has the potential to fight multidrug resistance of cancer. *BioMed Research International* 2015: 1–24.
62. Koné, W.M.A., Koffi, A.G., Bomisso, E.L. and Tra Bi, F.H., 2012. Ethnomedical study and iron content of some medicinal herbs used in traditional medicine in Cote d'Ivoire for the treatment of anaemia. *African Journal of Traditional, Complementary and Alternative Medicines* 9(1): 81–87.
63. Merel, H., Akpovi, A. and Maesen, A.L.L.G.V.D., 2004. Medicinal plants used to treat malaria in Southern Benin. *Economic Botany* 58: S239–S252.
64. Bla Kouakou, B., Trebissou, J.N.D., Bidie, A.P., Assi, Y.J., Zirihi-Guede, N. and Djaman, A.J., 2015. Étude ethnopharmacologique des plantes antipaludiques utilisées chez les Baoulé- N'Gban de Toumodi dans le Centre de la Côte d'Ivoire. *Journal of Applied Biosciences* 85: 7775–7783.
65. Guédé, N.Z., N'guessan, K., Dibié, T.E. and Grellier, P., 2010. Ethnopharmacological study of plants used to treat malaria, in traditional medicine, by Bete Populations of Issia (Côte d'Ivoire). *Journal of pharmaceutical Sciences and Research* 2(4): 216–227.

2 Chemistry and Biological Activities of Cameroonian *Hypericum* Species

Gabin Thierry M. Bitchagno
University of Dschang

Vaderament-A Nchiozem-Ngnitedem
University of Dschang
University of Nairobi

Sylvain Valere Sob
University of Maryland

Serge Alain Fobofou
Harvard Medical School
Technische Universität Braunschweig

CONTENTS

DOI: 10.1201/9780429506734-2

2.1   INTRODUCTION

The genus *Hypericum* L. belongs to the family Hypericaceae Juss. and has a long medicinal history in the traditional pharmacopeia. *Hypericum* plant species are popularly used around the world as principal or associate solutions toward various illnesses. The main scientific interests in these plants started with *H. perforatum* (commonly known as St. John's wort), which is used against depression [1].

The flowering tops of *H. perforatum* are also prepared as a decoction or infusion and taken orally for sedative or tonic purposes. They are also applied externally as a poultice or prepared as an oil infusion to treat sciatica and neuralgia and to speed wound healing [2]. *Hypericum* species other than *H. perforatum* have also been ethnomedicinally used throughout the world. In China, *H. monogynum* L. is used for the treatment of hepatitis, acute laryngopharyngitis, conjunctivitis and snakebites [3]. *H. uralum* Buch. -Ham. ex D. Don is used in the Yunnan Province of China for anti-inflammation applications and detoxification and to relieve itching. In Japan, *H. chinense* is used as a folk medicine for the treatment of sterility [4]. *H. scabrum* L. is among the most popular medicinal plants in Uzbekistan and is used against bladder, intestinal and heart diseases, rheumatism and cystitis. In Mexico, where *H. uliginosum* H.B.K. is used for the treatment of diarrhea, it is called "Tzotzil" and also "rabbit plant" [5]. *H. laricifolium* H.B.K., with common name "Romerillo", has been used in traditional Ecuadorian medicine as a diuretic and for provoking menstruation. In the folk medicine of Papua New Guinea, leaves of *H. papuanum* Ridl. are applied to treat sores [6]. In Turkey, *H. empetrifolium* Willd. is used against kidney stones and gastric ulcers and in Greece, as an anthelmintic and diuretic [6]. *Hypericum* species show interesting biological activities *in vitro*, and the compounds responsible for these effects were described. Extracts from *Hypericum* species exhibit antibacterial, anti-depressive, cytotoxic, anti-inflammatory and antiviral effects [3,5]. Chemical investigations of these extracts led to the isolation of acylphloroglucinols, xanthones, naphthodianthrones and flavonoids among others. Some of these metabolites were reported from African *Hypericum* species, especially those growing in the Cameroonian flora and represented by *H. roeperianum* (syn. *H. riparium*), *H. lanceolatum* (syn. *H. revolutum*) and *H. peplidifolium* [7]. This chapter reviews the chemistry and biological activities of extracts and compounds from Cameroonian *Hypericum* species (Figure 2.1).

FIGURE 2.1   Picture of *H. lanceolatum* from Cameroon.

2.2 SECONDARY METABOLITES FROM CAMEROONIAN *HYPERICUM* SPECIES

Compounds isolated from Cameroonian *Hypericum* species belong to diverse natural product classes including acylphloroglucinols, phenyl polyketide derivatives, xanthones, anthraquinones and benzophenones. They were isolated from *H. roeperianum* (syn. *H. riparium*), *H. lanceolatum* (syn. *H. revolutum*) and *H. peplidifolium* [6,8–17].

2.2.1 ACYLPHLOROGLUCINOLS

Acylphloroglucinols are derivatives of 1,3,5-trihydroxybenzene, a compound also known as phloroglucinol. Phloroglucinol-type compounds are characteristic of the Hypericaceae family, which contains the genus *Hypericum*. They are the main chemotaxonomic markers of the genus and exhibit interesting biological activities. The most reputed compound of this group, occurring in the *Hypericum* genus, is hyperforin for its considerable antibacterial activity. Because of their phenolic nature, acylphloroglucinols cannot be specifically identified from a mixture using colored test. Phloroglucinols are the most representative constituents of Cameroonian *Hypericum* species (**1–28**). In general, acylphloroglucinols have been reported to possess several biological activities such as antidepressant, anti-staphylococcal, antibacterial, anti-inflammatory and antiproliferative [18,19]. Acylphloroglucinols were extracted and isolated from Cameroonian *Hypericum* species using standard solvent extraction techniques and reversed phase HPLC [6].

2.2.2 POLYKETIDES

Phenyl polyketides have been described from members of the genus *Hypericum* and are intensively considered as one of the chemotaxonomic markers, alongside acylphloroglucinols. Phenyl polyketides are a large group of secondary metabolites known to possess remarkable diversity in their structures. They contain alternated carbonyl and methylene groups or are derived from precursors which contain such alternating groups. Polyketide derivatives are produced by plants and insects but are reputed compounds of almost all microorganisms (bacteria, fungi, protists, insects, mollusks and sponges) with anticancer, antifungal and antiparasitic properties [20]. The most important bacteria producing polyketides is *Streptomyces* genus. In fact, polyketides are generated by microorganisms to protect themselves from attack. Only few members (**29–38**) of this class have been reported from Cameroonian *Hypericum* and are derived from the same core bearing a phenyl group (Figure 2.2). These metabolites were extracted and isolated from Cameroonian *Hypericum* species using standard solvent extraction techniques and reversed phase HPLC [14] (Figure 2.3).

2.2.3 OTHER PHENOLIC COMPOUNDS

Phenolic compounds are ubiquitous in plants and are obtained from the shikimate pathway. All the Cameroonian *Hypericum* species biosynthesize phenolic compounds. The genus *Hypericum* produced a number of xanthones (**38–62**), coumarins

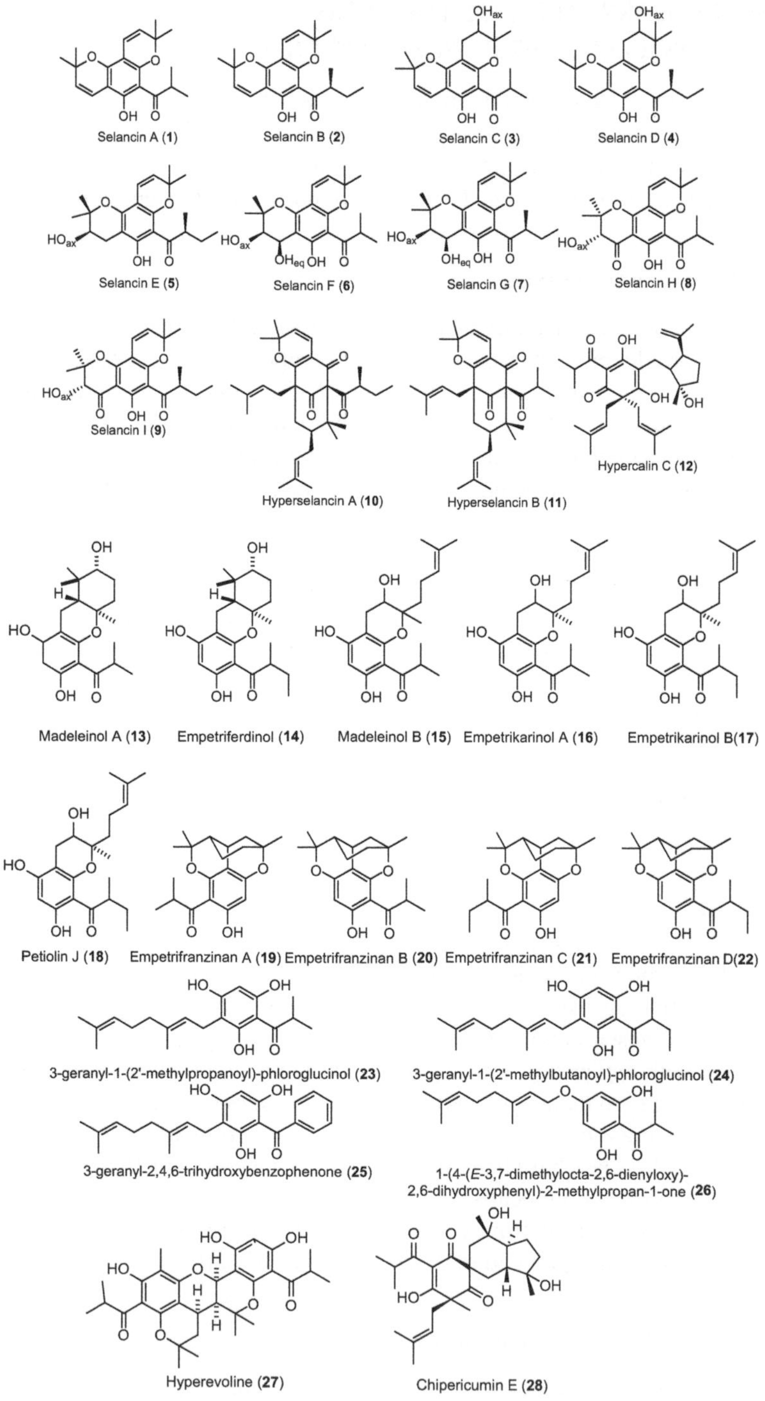

FIGURE 2.2 Structures of acylphloroglucinol derivatives from Cameroonian *Hypericum* species.

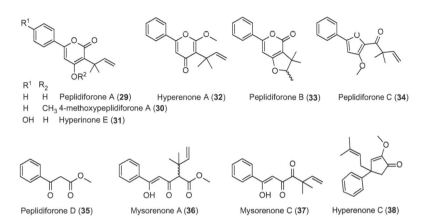

R¹ R²
H H Peplidiforone A (**29**)   Hyperenone A (**32**)   Peplidiforone B (**33**)   Peplidiforone C (**34**)
H CH₃ 4-methoxypeplidiforone A (**30**)
OH H Hyperinone E (**31**)

Peplidiforone D (**35**)   Mysorenone A (**36**)   Mysorenone C (**37**)   Hyperenone C (**38**)

FIGURE 2.3   Phenyl polyketides from Cameroonian *Hypericum* species.

(**63–67**), flavonoids (**68–71**), benzophenones (**72, 73**) and anthraquinones (**74, 75**). Xanthones are among the most abundant phenolic compounds found in Cameroonian *Hypericum* species, and their biological activity, which is related to their tricyclic nature, has been documented. Xanthones are known to exhibit a wide range of biological activities such antiproliferative, antimicrobial, anti-inflammatory, antioxidant and antihypertensive among others [21,22]. The mode of formation of xanthone has been reported by many authors. For instance, in lower plant species, the biosynthetic pathway of xanthone results from the condensation of eight units of malonyl-CoA, while in higher plants, the formation of xanthone involves the condensation of shikimate acid with three molecules of malonyl-CoA as an extended unit [23]. The origin of the other classes of phenolic compounds is related to that of xanthones. Mainly, whether coumarins, flavonoids, benzophenones or anthraquinones, phenolic compounds found their origin in the shikimate route. Most of them are antioxidants and exhibit interesting cytotoxic properties. Quinones are highly exploited to activate cysteine containing protein and enzyme, for example, caspase activation, to fight against cancer and related diseases [24] (Figures 2.4 and 2.5).

2.2.4 TRITERPENOIDS

Terpenoids are the largest class of secondary metabolites in the plant kingdom. Only a very small number of this class of compounds were isolated from Cameroonian *Hypericum* species: triterpenes (**76, 77**) and steroids (**78–83**). A literature survey showed that triterpenoids possess diverse pharmacological activities including antimicrobial, anti-inflammatory, cytotoxic, and enzyme inhibitory [25,26] (Figure 2.6).

2.3 UNIQUENESS OF CAMEROONIAN *HYPERICUM* SPECIES AND COMPOUNDS THEREFROM

Among roughly half of thousand species of *Hypericum* genus, only very few are found in Cameroon including *H. lanceolatum, H. peplidifolium, H. roeperianum* (syn. *H. riparium*), *H. lanceolatum* also known as *H. revolutum* subsp. revolutum,

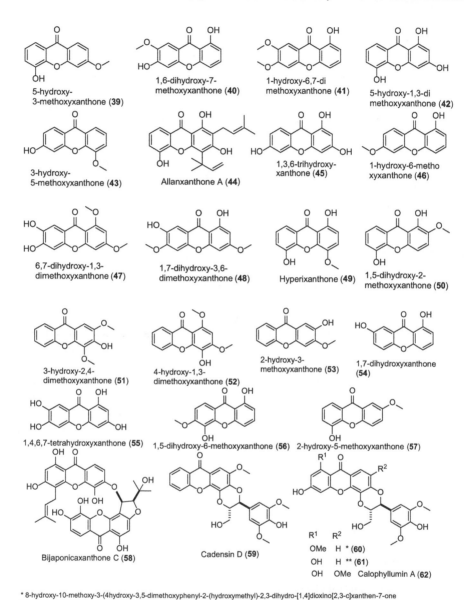

FIGURE 2.4   Xanthones from Cameroonian *Hypericum* species.

Norysca lanceolata, Campylosporus angustifolius, Campylosporus madagascariensis, Campylosporus reticulatus, H. angustifolium, H. madagascariense and *H. revolutum*. None of these species are solely endemic to Cameroon [12]. Nevertheless, *H. peplidifolium* can only be found in sub-Saharan regions of Africa including Cameroon. They are distributed in the mountain regions of the country, mainly in the West and South West regions [6,8–17].

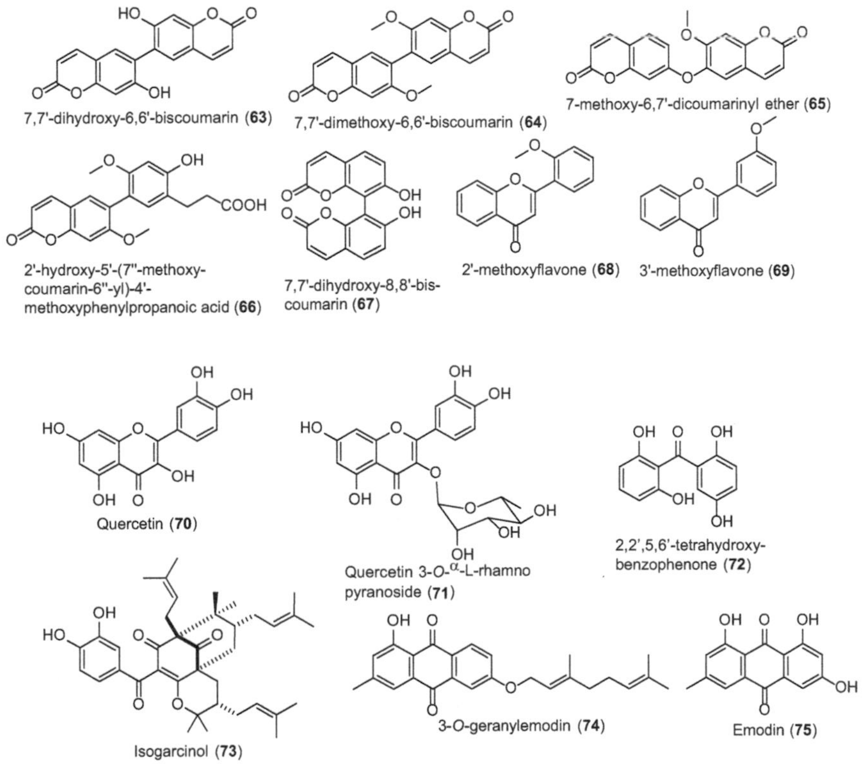

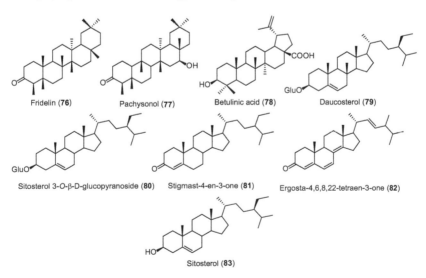

FIGURE 2.5 Biscoumarins (**63–67**), flavonoids (**68–71**), benzophenones (**72, 73**), and anthraquinones (**74, 75**) from Cameroonian *Hypericum* species.

FIGURE 2.6 Triterpenes from Cameroonian *Hypericum* species.

Cameroonian *Hypericum* species have produced valuable compounds with new skeletons. For instance, selacins H (**8**) and I (**9**) with a 6-acyl-2,2-dimethylchroman-4-one core are unprecedented in nature [13]. Likewise, the Cameroonian *H. riparium* afforded the only existing biscoumarins (**63–67**) described from the genus *Hypericum* so far [16].

2.4 BIOLOGICAL ACTIVITIES OF SECONDARY METABOLITES FROM CAMEROONIAN *HYPERICUM* SPECIES

The biological activity of fractions and compounds isolated from Cameroonian *Hypericum* species was described. The samples were tested for cytotoxic, antifungal, antioxidant, anthelmintic and antimicrobial activities. These are summarized in Table 2.1 below.

2.5 ETHNOMEDICINAL USES VERSUS BIOLOGICAL ACTIVITIES OF EXTRACTS AND COMPOUNDS FROM CAMEROONIAN *HYPERICUM* SPECIES

Hypericum lanceolatum or "Mekanaghene" in the Lebialem Sub-Region of Cameroon is a multipurpose plant, commonly used in Cameroon traditionally to treat several ailments including malaria, skin infections, venereal diseases, gastrointestinal disorders, tumors, infertility, and epilepsies [10]. Leaves of the plant are extracted with palm wine and used for the treatment of skin infections, epilepsy, and tumors, while the roots are boiled in water and used to treat venereal diseases, gastrointestinal disorders, and infertility. The stem bark is usually boiled in water and administered either as a steam bath or orally for the treatment of malaria and other fevers. The roots are also known for their use against intestinal worms and dysentery. They are combined with *Mangifera indica* leaves, boiled, and administered as a drink. In the Lebialem Division (South West Region), a decoction of fresh leaves is taken orally to treat nerve problems [12].

Zofou et al. have evaluated the antimalarial activity of extract, fractions, and compounds from *H. lanceolatum* against the reference strain W2mef (MRA-615) of *Plasmodium falciparum*, and the field isolated one SHF4 from a malaria patient in Buea in the South West region of Cameroon [12]. The cytotoxicity profiles of the extracts and pure compounds were evaluated against LLC-MK2 monkey kidney epithelial cells. The work aimed to validate the traditional claims of the plant to relieve malaria in the region. Following a bioassay-guided fractionation of the crude methanol extract of the stem back of this species, a subsequent EtOAc subfraction led to the major fraction F3, the most active part of the plant with $IC_{50} < 4$ µg/mL. Consequently, the purification of fraction F3 led to the isolation of three compounds with similar activities. Among them, 5-hydroxy-3-methoxyxanthone (**39**) was the most active constituent ($IC_{50} < 3$ µg/mL).

H. roeperianum is used in Cameroonian traditional medicine against worm infestations. The crude chloroform extract of this plant exhibited significant anthelmintic activity against the model organism *Caenorhabditis elegans* with

TABLE 2.1

Biological Activities of Secondary Metabolites from Cameroonian *Hypericum* Species

Samples Tested	*Hypericum* species	Bioactivity	Reference
Cytotoxic activity			
Selancin A (**1**)	*H. lanceolatum*	No significant growth inhibition was observed while tested against colon (HT-29) and prostate (PC-3) cancer cell lines at concentration 10 μM	[13]
Selancin B (**2**)			
Selancin C (**3**)			
Selancin D (**4**)			
Selancin E (**5**)			
Selancin F (**6**)			
Selancin H (**8**)			
Selancin I (**9**)			
3-*O*-geranylemodin (**74**)			
Methanol extract		At 50 μg/mL, the methanol extract showed significant growth inhibition (88.6 ± 0.6%) in PC-3 versus (88.7 ± 1.8%) against HT-29 cancer cell line. The same extract at 500 μg/mL did not exhibit any anthelmintic activity	
Betulinic acid (**78**)	*H. lanceolatum*	Showed cytotoxicity against rhesus monkey kidney epithelial cells (LLC-MK2 Line) with CC_{50} = 25 μg/mL	[12]
Chloroform extract	*H. roeperianum*	The $CHCl_3$ extract showed cell viability of 16% and 18% against HT-29 and PC-3, respectively, at 50 μg/mL. However, no inhibition was observed at 0.5 μg/mL	[6]
Madeleinol A (**13**)	*H. roeperianum*	No cytotoxicity against PC-3 cancer line at 10 μM	
Madeleinol B (**15**)		No cytotoxicity against PC-3 cancer line at 10 μM	
Empetrikarinol B (**17**)		Weak cytotoxicity against HT-29 and inactive against PC-3 cancer line at 10 μM	
3-geranyl-2,4,6-trihydroxybenzophenone (**25**)		No cytotoxicity against PC-3 cancer line at 10 μM	
3-geranyl-1-(2′-methylpropanoyl)-phloroglucinol (**23**)		Weak cytotoxicity against PC-3 cancer line at 10 μM	

(Continued)

TABLE 2.1 (Continued)
Biological Activities of Secondary Metabolites from Cameroonian Hypericum Species

Samples Tested	Hypericum species	Bioactivity	Reference
3-geranyl-1-(2'-methylbutanoyl)-phloroglucinol (24)		Weak cytotoxicity against PC-3 and HT-29 cancer line at 10 μM	
Empetrifranzinan A (19)		Weak cytotoxicity against PC-3 cancer line at 10 μM	
Empetrifranzinan B (20)		Weak cytotoxicity against PC-3 and HT-29 cancer lines at 10 μM	
Empetrifranzinan C (21)		Weak cytotoxicity against both cancer cell lines	
Empetrifranzinan D (22)		No cytotoxicity observed against PC-3 cancer line at 10 μM	
Methanol extract	H. peplidifolium	Growth inhibition of 54% and 48% against HT-29 and PC-3C, respectively, at 50 μg/mL and was not active at 0.5 μg/mL	[14]
Ethyl acetate extract		Cell viability of 65% (HT-29) and 64% (PC-3) at 50 μg/mL, and no activity was observed at 0.5 μg/mL	
Peplidiforone A (29)	H. peplidifolium	At a concentration of 10 μM, all isolated compounds were inactive (cell viability > 85%)	
Peplidiforone C (34)			
4-methoxypeplidiforone A (30)			
Mysorenone-A (36)			
Mysorenone-C (37)			
Petiolin J (18)			
7,7'-dihydroxy-8,8'-biscoumarin (67)	H. riparium	The isolated compounds did not exhibit any significant cytotoxic activity against human prostate (PC-3) and colon (HT-29) cancer cell line	[16]
7-methoxy-6,7'-dicoumarinyl ether (65)			
2'-hydroxy-5'-(7''-methoxycoumarin-6''-yl)-4'-methoxyphenylpropanoic acid (66)			
7,7'-dimethoxy-6,6'-biscoumarin (64)			
CH2Cl2/MeOH extract	H. riparium	Cytotoxic IC50 = 9.74 μg/mL	[8]
Hyperenone A (32)	H. riparium	Cytotoxic against the human gastric (BGC-823) cell line with the IC50 = 17.64 μM	
1,5-Dihydroxy-2-methoxyxanthone (50)	H. riparium	Cytotoxic (BGC-823, IC50 = 18.50 μM)	

(Continued)

TABLE 2.1 (Continued)
Biological Activities of Secondary Metabolites from Cameroonian *Hypericum* Species

Samples Tested	*Hypericum* species	Bioactivity	Reference
Infection disease			
5-hydroxy-3-methoxyxanthone (39)	*H. riparium*	Antibacterial activity MIC = 0.97–250 µg/mL	[9]
Cadensin D (59)		Antifungal activity MIC = 0.97–250 µg/mL	
Ethyl acetate fraction		The EtOAc extract showed good antibacterial and antifungal activities against the tested organisms compared to the crude CH_2Cl_2/MeOH	
5-hydroxy-3-methoxyxanthone (39)	*H. lanceolatum*	Anti-plasmodial IC_{50} <3.50 µM	[12]
Betulinic acid (78)		Anti-plasmodial IC_{50} <5 µM	
EtOAc fraction		Good anti-plasmodial activity was observed against W2mef strain of *P. falciparum*, (IC_{50} <10 µg/mL)	
2,2',5,6'-tetrahydroxy benzophenone (72)	*H. lanceolatum*	Antibacterial MIC = 2–8 µg/mL	[11]
Isogarcinol (73)		Antioxidant RSa_{50} = 1.012 ± 0.247 µg/mL	
3-geranyl-1-(2'-methylbutanoyl)-phloroglucinol (24)	*H. roeperianum*	Anthelmintic of 37% ± 12% compared to the standard thiabendazole 19 ± 11%	[6]
Chloroform extract	*H. roeperianum*	The extract displayed a nematode death percentage against *Caenorhabditis elegans* of 90%–92%	
Methanol extract	*H. peplidifolium*	Antifungal activity	[15]
Ethyl acetate extract			
Peplidiforone A (29)	*H. peplidifolium*	Antifungal activity with growth inhibition zone of 34%–47% at 83.3 µM	
4-methoxypeplidiforone A (30)			
Dichloromethane/Methanol extract	*H. riparium*	Antimicrobial activity IC_{50} = 48.05 µg/mL	[8]
Chipericumin E (28)		Antimicrobial against *S. Aureus* (6.54 µg/mL ≤ IC_{50} ≤ 16.90 µg/mL)	
Hypercalin C (12)			
Hyperenone C (38)			
Hyperenone A (32)			
1,5-Dihydroxy-2-methoxyxanthone (50)			
Emodin (75)			

a percentage of nematode death higher than 80% at 500 µg/mL. The chemical investigation of this extract afforded the major compounds **23** and **24**, which possess anthelmintic activity [6].

2.6 CONCLUSION

Few members of the genus *Hypericum* are present in Cameroon. The Cameroonian species mostly represented by *H. lanceolatum, H. roeperianum,* and *H. peplidifolium* have been investigated both chemically and pharmacologically. Some of the biological activities supported the uses of these plant species in the Cameroonian traditional medicine. Chemical investigation of extracts afforded several novel compounds, some of them being unique to Cameroonian *Hypericum* species.

REFERENCES

1. Galeotti, N., 2017. *Hypericum perforatum* (St John's wort) beyond depression: A therapeutic perspective for pain conditions. *J. Ethnopharmacol.* 200: 136–146.
2. Crockett, S.L. and Robson, N.K., 2011. Taxonomy and chemotaxonomy of the genus *Hypericum. Med. Aromat. Plant Sci. Biotechnol.* 5: 1–13.
3. Xu, W.-J., Zhu, M.-D., Wang, X.-B., Yang, M.-H., Luo, J. and Kong, L.-Y., 2015. Hypermongones A-J, rare methylated polycyclic prenylated acylphloroglucinols from the flowers of *Hypericum monogynum. J. Nat. Prod.* 78: 1093–1100.
4. Tanaka, N. and Takaishi, Y., 2006. Xanthones from *Hypericum chinense. Phytochemistry* 67: 2146–2151.
5. Parker, W.L. and Johnson, F., 1968. The structure determination of antibiotic compounds from *Hypericum uliginosum. J. Am. Chem. Soc.* 90: 4716–4723.
6. Fobofou, S.A.T., Franke, K., Sanna, G., Porzel, A., Bullita, E., La Colla, P. and Wessjohann, L.A., 2015a. Isolation and anticancer, anthelmintic, and antiviral (HIV) activity of acylphloroglucinols, and regioselective synthesis of empetrifranzinans from *Hypericum roeperianum. Bioorg. Med. Chem.* 23: 6327–6334.
7. Hutchinson, J. and Dalziel, J.M., 1954. Flora of West Tropical Africa. Ed. Keay, R.W., *Crown agents for overseas governments and administrations,* Millbanks, London S.W.1 (U.K.), p. 295.
8. Tala, M.F., Talontsi, F.M., Zeng, G-Z., Wabo, H.K., Tan N-H., Spiteller, M. and Tane, P., 2015. Antimicrobial and cytotoxic constituents from native Cameroonian medicinal plant *Hypericum riparium. Fitoterapia* 102: 149–155.
9. Tala, M.F., Tchakam, P.D., Wabo, H.K., Talontsi, F.M., Tane, P., Kuiate, J-R., Tapondjou, L.A. and Laatsch, H., 2013. Chemical constituents, antimicrobial and cytotoxic activities of *Hypericum riparium* (Guttiferae). *Rec. Nat. Prod.* 7: 65–68.
10. Wabo, H.K., Kowa, T.K., Lonfouo, A.H.N., Tchinda, A.T., Tane, P., Kikuchi, H. and Frédérich M., 2012. Phenolic Compounds and Terpenoids from *Hypericum lanceolatum. Rec. Nat. Prod.* 6: 94–100.
11. Tchakam, P.D., Lunga, P.K., Kowa, T.K., Lonfouo, A.H.N., Wabo, H.K., Tapondjou, L.A., Tane, P. and Kuiat, J-R., 2012. Antimicrobial and antioxidant activities of the extracts and compounds from the leaves of *Psorospermum aurantiacum* Engl. and *Hypericum lanceolatum* Lam. *BMC Complementary Altern. Med.* 12: 136.
12. Zofou, D., Kowa, T.K., Wabo, H.K., Ngemenya, M.N., Tane, P. and Titanji, V.P.K., 2011. *Hypericum lanceolatum* (Hypericaceae) as a potential source of new anti-malarial agents: A bioassay-guided fractionation of the stem bark. *Malar. J.* 10: 167.

13. Fobofou, S.A.T., Franke, K., Porzel, A., Brandt, W. and Wessjohann, L.A., 2016a. Tricyclic acylphloroglucinols from *Hypericum lanceolatum* and regioselective synthesis of selancins A and B. *J. Nat. Prod.* 79: 743–753.

14. Fobofou, S.A.T., Harmon, C.R., Lonfouo, A.H., Franke, K., Wright, S.T. and Wessjohann, A.L., 2016b. Prenylated phenyl polyketides and acylphloroglucinols from *Hypericum peplidifolium*. *Phytochemistry* 124: 108–113.

15. Fobofou, S.A.T., Franke, K., Schmidt, J. and Wessjohann, L.A., 2015b. Chemical constituents of *Psorospermum densipunctatum* (Hypericaceae). *Biochem. Syst. Ecol.* 59: 174–176.

16. Fobofou, S.A.T., Franke, K., Arnold, N., Schmidt, J., Wabo, H.K., Tane, P. and Wessjohann, L.A., 2014. Rare biscoumarin derivatives and flavonoids from *Hypericum riparium*. *Phytochemistry* 105: 171–177.

17. Damen, F., Demgne, O.M.F., Bitchagno, G.T.M., Celik, I., Mpetga, J.D.S., Tankeo, S.B., Opatz, T., Kuete, V. and Tane, P., 2019. A new polyketide from the bark of *Hypericum roeperianum* Schimp. (Hypericaceae). *Nat. Prod. Res.* doi.org/10.1080/14786419.2019. 1677655.

18. Coqueiro, A., Choi, Y.H., Verpoorte, R., Gupta, K.B.S.S., Mieri, M.D., Hamburger, M., Young, M.C.M., Stapleton, P., Gibbons, S. and Bolzani, V. da S., 2016. Antistaphylococcal prenylated acylphorglucinol and xanthones from *Kielmeyera variabilis*. *J. Nat. Prod.* 79: 470–476.

19. Qiu, D., Zhou, M., Chen, J., Wang, G., Lin, T., Huang, Y., Yu, F., Ding, R., Sun, C., Tian, W. and Chen, H., 2020. Hyperelodiones A-C, monoterpenoid polyprenylated acylphorglucinols from *Hypericum elodeoides*, induce cancer cells apoptosis by targeting RXRα. *Phytochemisry* 170: 112216.

20. Huffman, J., Gerber, R. and Du, L., 2010. Recent advancements in the biosynthetic mechanisms for polyketide-derived mycotoxins. *Biopolymers* 93: 764–776.

21. Ahmed, H.A., Bozzi, B., Correa, M., Capson, T.L., Kursar, T.A., Coley, P.D., Solis, P.N. and Gupta, M.P., 2003. Bioactive constituents from three *Vismia* species. *J. Nat. Prod.* 66: 858–860.

22. Nawong, B., Karalai, C., Chantrapromma, S., Ponglimanont, C., Kanjana-Opas, A., Chantrapromma, K. and Fun, H-K., 2007. Quinonoids from the barks of *Cratoxylum formosum* subsp. *Pruniflorum*. *Canadian J. Chem.* 85: 341–345.

23. Masters, K-S. and Brase, S., 2012. Xanthones from fungi, lichens, and bacteria: The natural products and their synthesis. *Chem. Rev.* 112: 371776.

24. Li, W.W., Heinze, J. and Haehnel, W., 2005. Site-specific binding of quinones to proteins through thiol addition and addition-elimination reactions. *J. Am. Chem. Soc.* 127: 6140–6141.

25. Djoumessi, A.V.B., Sandjo, L.P., Liermann, J.C., Schollmeyer, D., Kuete, V., Rincheval, V., Berhanu, A.M., Yeboah, S.O., Wafo, P., Ngadjui, B.T. and Opatz, T., 2012. Donellanic acids AC: New cyclopropanic oleanane derivatives from *Donella ubanguiensis* (Sapotaceae). *Tetrahedron* 68: 4621–462.

26. Christophe, L., Beck, J., Cantagrel, F., Marcourt, L., Vendier, L., David, B., Plisson, F., Derguini, F., Vandenberghe, I., Aussagues, Y., Ausseil, F., Lavaud, C., Sautel, F. and Massiot, G., 2012. Proteasome inhibitors from *Neoboutonia melleri*. *J. Nat. Prod.* 75: 34–47.

3 Biologically Active Saponins from Some Cameroonian Medicinal Plants

Léon Azefack Tapondjou,
Rémy Bertrand Teponno,
Beaudelaire Kemvoufo Ponou, and
Romuald Tematio Fouedjou
University of Dschang

Luciano Barboni
University of Camerino

CONTENTS

DOI: 10.1201/9780429506734-3

3.1   INTRODUCTION

Many research units in Cameroon are focused on the phytochemical and pharmacological studies of Cameroonian medicinal plants and their main objectives are based on the isolation, characterization, and biological screening of secondary metabolites from selected medicinal plants, followed by establishment of the link between the chemical composition of these plants and their traditional uses in folk medicine. In particular, the saponins have been studied intensively. Saponins are naturally occurring sugar conjugates of triterpene or steroid aglycones. Owing to their biological activities, they have been used for a long time both in ethnic as well as in conventional medicine. This chapter describes 195 saponins including about 115 new ones, which have been isolated during the last 15 years from Cameroonian medicinal plants and classified based on their different families. Several studies carried out on saponins show that they possess biological activities.

3.2   SAPONINS

Saponins are steroid or triterpene glycoside compounds widely distributed in the plant and marine animal kingdoms. The term saponin derives from the Latin word *sapo* (soap) and refers to its capacity to form persistent foam when shaken with water, even in dilute solution. This easily observable character has attracted human interest from ancient times. The classical definition of saponins is then based on their surface activity; many of them have detergent properties [1]. *Saponaria officinalis* (common name soapwort) is an example of a saponin containing plant that has been employed for hundreds of years as a natural soap [1]. Saponins also possess the property of forming stable froth when shaken with water, show hemolytic activity, have a bitter taste, and are toxic to fish. These attributes while not common to all saponins have frequently been used to characterize this class of secondary metabolites. Their properties are correlated with the amphiphilic character of the molecule (lipophilic and hydrophilic moieties).

The aglycone or nonsaccharide portion of the saponin molecule is called the genin or sapogenin, or sometimes both terms are used in the same article. Depending on the type of sapogenins or genins present, the saponins can be divided into three major classes [2,3]. The first group consists of the steroidal saponins, which are almost exclusively present in the monocotyledonous angiosperms. The second group consists of the triterpenoid saponins, which are the most common and occur mainly in the dicotyledonous angiosperms [4]. The third group is called steroidal amines, which are classified by others as steroidal alkaloids [4]. The glycine parts of these compounds are generally connected to oligosaccharides linear or branched, attached through a hydroxyl or a carboxyl group or both. The sites of attachment may be one (monodesmosides), two (bidesmosides), or three (tridesmosides). The most common monosaccharide moieties found in saponins are glucose, galactose, mannose, rhamnose, fucose, xylose, arabinose, apiose, etc. Several *in vitro* and *in vivo* tests on individual saponins have revealed a broad spectrum of biological and pharmacological activities that have been reported in several reviews. Owing to their biological activities, they have been used for a long time both in ethnic as well as in conventional medicine where they have been employed extensively in the drug-related industry, vaccine adjuvants, food industries, and pharmaceutical products. The role of saponins in the producer organisms has not been completely clarified, but their action in plant defense has been reported in the literature [5].

This chapter reports and summarizes research on saponins isolated from different parts of some plants used in traditional medicine in Cameroon. The biological effects (cytotoxic, antimicrobial, anti-inflammatory, antioxidant, and antiproliferative activities) of these compounds are also reviewed.

3.2.1 Biosynthesis of Saponins

The terpenoids form a large and structurally diverse family of natural products derived from C_5 isoprene units joined in a head-to-tail fashion. Two pathways are leading to isoprene; the mevalonate pathway and the recently discovered mevalonate-independent pathway via deoxyxylulose phosphate [6]. The process of the biosynthesis of the saponin genins proceeds via the isoprenoid pathway in which three isoprene units are first linked in a head-to-tail manner to each other, resulting in a 15-carbon-atom molecule farnesyl pyrophosphate. Two farnesyl pyrophosphates are subsequently linked in a tail-to-tail manner to give a compound of 30 carbon atoms, called squalene (Scheme 3.1) [7].

Squalene is oxidized to oxidosqualene which is converted to cyclic derivatives via protonation and epoxide ring opening leading to a carbocation that can undergoes several types of cyclization reactions [7]. Cyclization of squalene is via the intermediate squalene-2,3-oxide. The main cyclization and rearrangement reactions leading to the triterpenoid and steroid skeletons are shown in Scheme 3.2. The aglycones are then tailored with oxidoreductases before being glycosylated with multiple sugar moieties [8].

In the process of biosynthesis of terpenoids, the production of the intermediate depends on the conformation of the squalene oxide within the enzyme: the cyclization of the chair-chair-chair conformation forms the dammarenyl carbocation

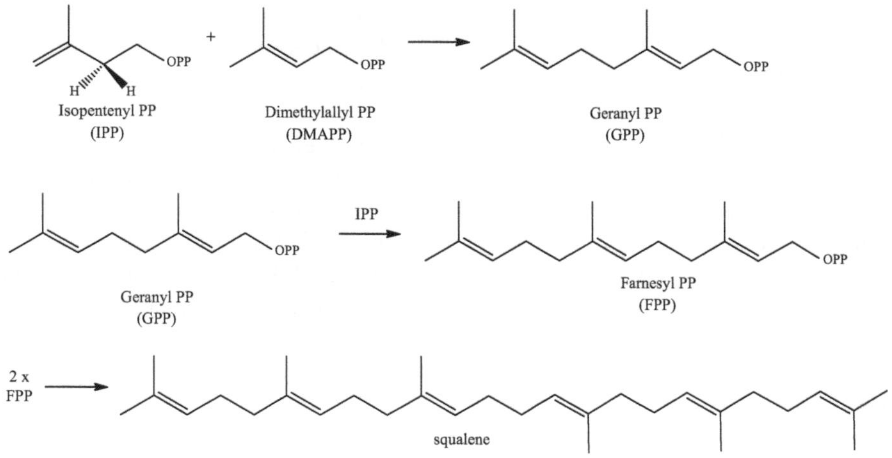

SCHEME 3.1   Biosynthesis of squalene.

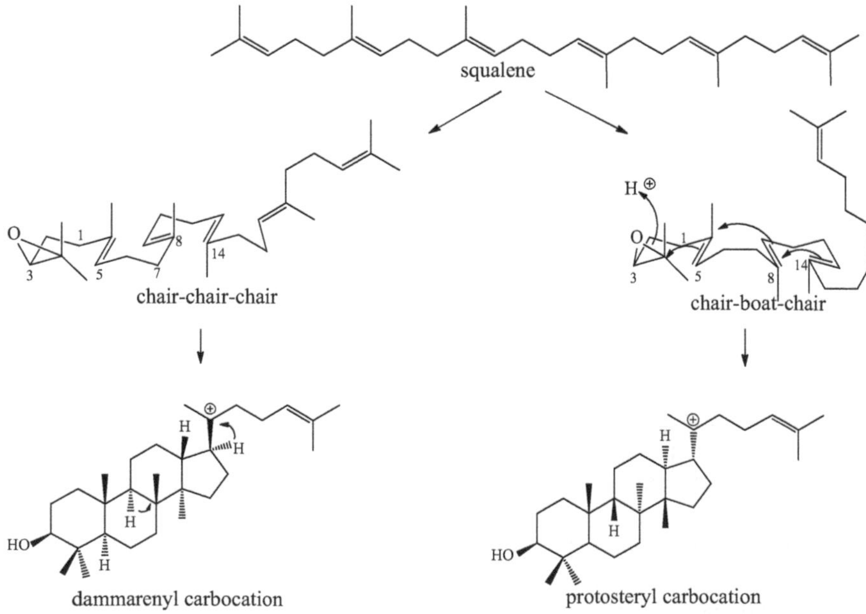

SCHEME 3.2   Biosynthesis of dammarenyl and protosteryl carbocations from squalene oxide.

leading to dammarane, ursane, oleanane, lupane, or taraxasterane type saponins, while the cyclization of the chair-boat-chair conformation of the squalene produces the protosteryl carbocation leading to cycloartane, lanostane, cucurbitane, or steroidal type saponins, such as cholestane, furostane, and spirostanes as indicated in Schemes 3.2 and 3.3.

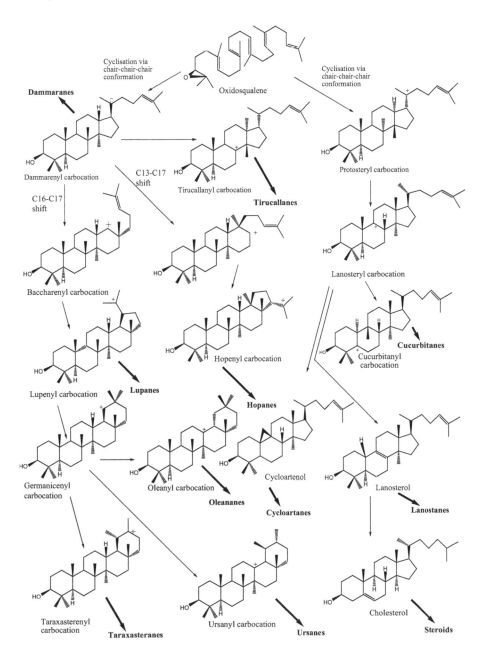

SCHEME 3.3   Cyclization of oxidosqualene to the various aglycone saponin skeletons.

It was previously considered that saponins of the spirostanol type may be artefacts formed from furostanol glycosides by hydrolysis during isolation [9]. In the natural state, the sugar linked to the C-26 hydroxyl group protects the furostanol saponins from ring closure, but rupture of the C-26 glycosidic bond then causes the formation of spirostanol glycosides. Recent studies showed that furostanol and spirostanol are

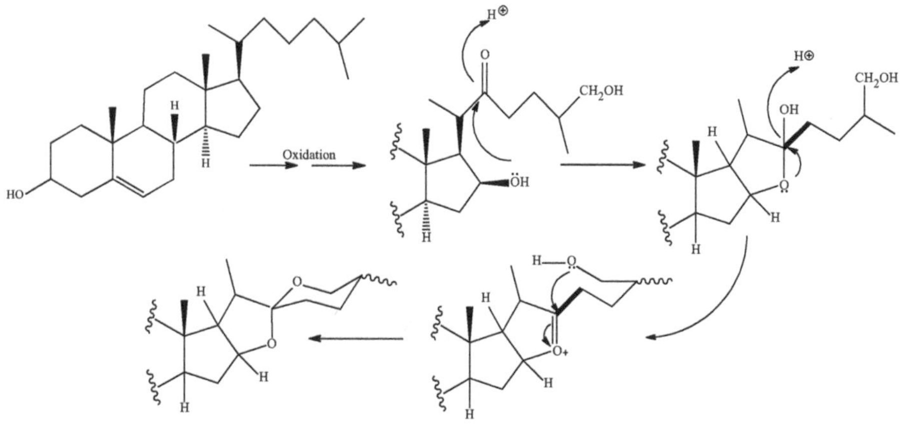

SCHEME 3.4   The biosynthesis of spiroskeletal compounds.

C-27 sterols in which the cholesterol sidechain has been modified. This occurs by a series of oxygenation reactions in which C-16 and one of the side-chain terminal methyl groups are hydroxylated and C-22 is oxidized to a ketone. The intermediate is converted into the hemiketal and then the spiroketal forms (Scheme 3.4) [10].

3.2.2   ISOLATION AND PURIFICATION OF SAPONINS

Isolation of saponins demands tedious and sophisticated techniques because of their unique chemical nature. Their isolation from plant material is complicated also by the occurrence of many closely related substances in plant tissues and by the fact that most of the saponins lack a chromophore. The modern methods available for the separation and analysis of saponins have been well reviewed [11–13].

There are several strategies available for their isolation. As a general rule, this begins with the extraction of the plant material with aqueous methanol, methanol, or ethanol. Further processing of the extract is carried out after evaporation under reduced pressure, dissolution in a small amount of water and phase separation into *n*-butanol. It is currently recognized that this step is sometimes undesirable, since only those saponins with short oligosaccharide side chains will eventually be extracted into the butanolic phase. A further purification is then carried out, which involves liquid chromatography over a silica gel column, or a gradient elution from a polymeric support or liquid-liquid partition chromatography, or the most commonly employed HPLC and MPLC separations. In most cases, several steps have to be repeated with a change of support or eluent to achieve high purity. The solvent systems generally used include $CHCl_3$-MeOH-H_2O and EtOAc-MeOH-H_2O on normal phase or MeOH-H_2O and MeCN-H_2O on reverse phase [14].

3.2.3   STRUCTURAL ELUCIDATION

Once the saponin has been purified, it is usually subjected to a complex structural elucidation process, which is based on the determinations of the following:

- the aglycone structure,
- the composition and sequence of the monosaccharides in the carbohydrate moiety,
- the location of linkages between monosaccharide units,
- the anomeric configuration of each monosaccharide unit,
- the location of the carbohydrate moiety on the aglycone

The above structural elucidation steps are mainly based on NMR spectroscopic analysis and integrated with MS and infrared spectroscopy.

3.2.4   STRUCTURAL ELUCIDATION OF THE AGLYCONE PORTION

The structural elucidation of the aglycone portion is based on the comparison of ^{1}H and ^{13}C NMR data with those published. Due to the complexity of NMR spectra of saponins, when differences with data reported in the literature are encountered, then extensive 2D NMR studies are needed for the determination of substitution sites as well as stereochemical assignments. Further, because of the easy glycosidic cleavage, MS fragmentations can provide information on the molecular mass of the aglycone portion.

For steroidal saponins, the multiplicity and chemical shift of the C-22 acts as a structural marker for the determination of the parental skeleton of the aglycone [15]. Also, triterpenoid skeletons may be recognized based on their ^{13}C NMR data. Many steroidal and triterpenoidal sapogenins are characterized by hydroxylation at a number of positions. Some spectral features, like the variation of chemical shift induced by the substituents, can be used to infer the presence of substituents. Extensive use of 2D NMR, usually, COSY and TOCSY experiments help determine the spin systems, while HMBC allows for connecting the partial substructures.

3.2.5   STRUCTURAL ELUCIDATION OF THE SUGAR PORTION

Due to the highly crowded region of the saccharide, the attribution of proton and carbon resonances is not easy, and a combination of 2D NMR experiments becomes very useful: COSY, HSQC, NOESY, or ROESY experiments and TOCSY experiments. Evaluation of the coupling constants is needed to identify the anomeric configuration. Small coupling constants (1–3 Hz) are observed for the α anomers, while large coupling constants are observed for the β anomers (7–8 Hz) of the pyranose sugars [16].

3.2.6   GLYCOSYLATION SITE AND SUGAR CHAINS BRANCHING

Once the structural units have been identified, they need to be connected. HMBC correlations confirm these points of glycosylation. Concerning the sugar chains, the interglycosidic linkages can be achieved on HMBC correlations usually observed among the anomeric proton and the carbon of the second sugar involved in the linkage. The ROESY and TOCSY experiments establish sugar chain interglycosidic linkages. In MS spectra, glycosidic cleavage can provide information on sequence and branching of sugar units. Over the past few years, coupled with NMR, MS has supported the structural elucidation of uncharacterized saponins, providing information about their elemental composition and/or structural insights related to their fragmentation patterns [17].

3.3 PHYTOCHEMICAL ASSESSMENT

3.3.1 TRITERPENOIDAL SAPONINS

3.3.1.1 Saponins from the Rubiaceae

The Rubiaceae are trees, shrubs, or infrequently herbs comprising about 450 genera and 6500 species. Economically important products of the family Rubiaceae include quinine which is derived from the bark of *Cinchona* species; coffee, from the seeds of *Coffea* species; ipecac, obtained from the roots of *Psychotria ipecacuanha* and also *Crossopteryx febrifuga* [1].

Mitragyna stipulosa is a tree widely distributed throughout Africa. Stem barks of some *Mitragyna* species have been used to treat cough and malaria and as a diuretic by the local people of West Africa and Congo for a long time. The stem bark has been used to treat neuralgia pain, cancer, and diabetes by the local people of southern Cameroon for a long time [18]. A phytochemical study of an ethanolic extract of barks of *Mitragyna stipulosa* led to isolation of quinovic acid glycoside derivatives: their structures were identified as quinovic acid 3-*O*-[β-D-glucopyranoside] (quinovin glycoside C) (**1**) and quinovic acid 3-*O*-[β-D-quinovopyranoside]-27-*O*-[β-D-glucopyranosyl] ester (**2**) [19].

The genus *Oxyanthus* consists of more than 40 species distributed throughout the African continent, south of the Sahara. *Oxyanthus pallidus* is a small shrub widespread from Senegal to Nigeria, extending from Sudan to Ethiopia. This plant is extensively used in the west region of Cameroon as an analgesic [20]. From the MeOH extract of leaves of *Oxyanthus pallidus*, three cycloartane glycosides, named pallidiosides A-C were isolated from repeated chromatographic purification over silica gel column. The structures of pallidioside A (**3**), B (**4**), and C (**5**) were elucidated to be 7-*O*-β-D-glucopyranosylcycloart-3-β, 16-β, 25-trihydroxy-24-one; 7-*O*-β-D-glucopyranosylcycloart-3β, 16β, 24(*S*), 25-tetraol and 7-*O*-β-D-glucopyranosylcycloart-3β, 16β, 24(*R*), 25-tetraol respectively [21].

Crossopteryx febrifuga is a deciduous tree widely used in Northern Nigeria in the therapeutic management of trypanosomiasis, malaria, dysentery, diarrhea, fevers, and painful inflammatory disorders [22]. Phytochemical investigation of the stem bark extract of *Crossopteryx febrifuga* resulted in the isolation of 18-epi-3β-D-glucopyranosylurs-12, 20(30)-diene-27,28-dioïc acid (**6**) [23] (Figure 3.1).

3.3.1.2 Saponins from Mimosaceae

The Mimosaceae are mostly tropical and subtropical trees and shrubs comprising about 64 genera and 2000 species. This family of spiny woody plants, commonly included in the family Leguminosae comprises *Entada rheedii*, *Entada pursaetha*, *Albizia coriaria*, *Albizia zygia*, *Albizia glaberrima*, *Albizia lebbeck*, *Albizia*

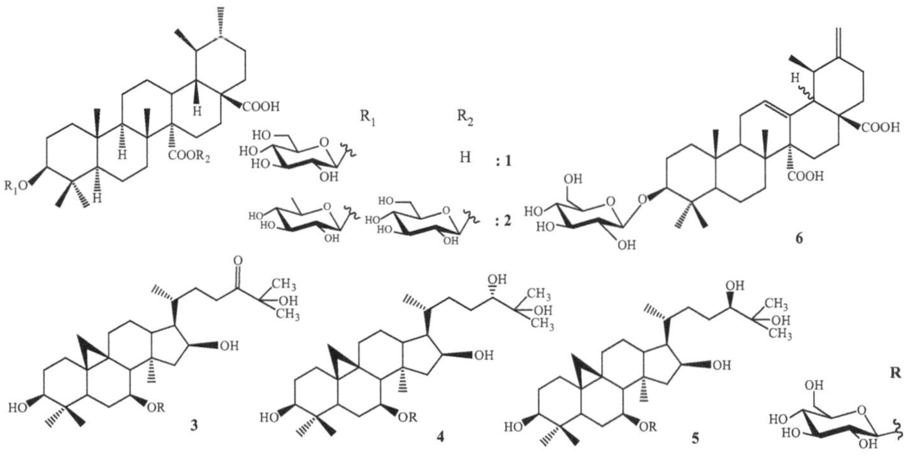

FIGURE 3.1 Saponins from *Mitragyna stipulosa*, *Oxyanthus pallidus* and *Crossopteryx febrifuga.*

boromoensis, *Piptadeniastrum africanum*, *Cylicodiscus gabunensis*, *Acacia albida*, *Tetrapleura tetraptera* species [24,25].

Five new triterpenoid saponins, pursaethosides A-E (**7–11**) elucidated as 3-*O*-β-D-xylopyranosyl-(1→3)-[α-L-arabinopyranosyl-(1→6)]-2-acetylamino-2-deoxy-β-D-glucopyranosyloleanolic acid 28-*O*-β-D-apiofuranosyl-(1→2)-β-D-glucopyranos ide (pursaethoside A) (**7**), 3-*O*-β-D-xylopyranosyl-(1→3)-*O*-[α-L-arabinopyranosyl-(1→6)]-2-acetylamino-2-deoxy-β-D-glucopyranosylechinocystic acid 28-*O*-β-D-apio furanosyl-(1→2)-β-D-glucopyranoside (pursaethoside B) (**8**), 3-*O*-β-D-xylopyranos yl-(1→3)-*O*-[α-L-arabinopyranosyl-(1→6)]-2-acetylamino-2-deoxy-β-D-glucopyranosylentagenic acid 28-*O*-β-D-apiofuranosyl-(1→2)-β-D-glucopyranoside (pursaethoside C) (**9**), 3-*O*-β-D-xylopyranosyl-(1→3)-[α-L-arabinopyranosyl-(1→6)]-2-acety-lamino-2-deoxy-β-D-glucopyranosylentagenic acid 28-O-β-D-apiofuranosyl-(1→2)-[β-D-apiofuranosyl-(1→4)]-β-D-glucopyranoside (pursaethoside D) (10) and 3-O-β-D-xylopyranosyl-(1→3)-[α-L-arabinopyranosyl-(1→6)]-2-acetylamino-2-deoxy-β-D-glucopyranosylentagenic acid 28-*O*-β-D-apiofuranosyl-(1→2)-[β-D-apiofuranosyl-(1→3)]-[β-D-glucopyranosyl(1→4)]-β-D-glucopyranoside (pursaethoside E) (**11**) were isolated from the *n*-BuOH extract of the kernel seeds of *Entada pursaetha* [26].

From the stem bark of *Tetrapleura tetraptera*, two previously unreported ole-anane-type saponins, 3-*O*-{6-*O*-[(2*E*, 6*S*)-2,6-dimethyl-6-hydroxyocta-2,7-dienoyl]-β-D-glucopyranosyl-(1→2)-β-D-glucopyranosyl-(1→3)-β-D-glucopyranosyl-(1→4)-[β-D-glucopyranosyl-(1→2)]-β-D-glucopyranosyl}-3,27-dihydroxyoleanolic acid (tetrapteroside A) (**12**) and 3-*O*-{β-D-glucopyranosyl-(1→ 2)-6-*O*-[(*E*)-feruloyl]-β-D-glucopyranosyl-(1→3)-β-D-glucopyranosyl-(1→4)-[β-D-glucopyranosyl-(1→2)]-β-D-glucopyranosyl}-3,27-dihydroxyoleanolic acid (tetrapteroside B) (**13**), were isolated [27] (Figure 3.2).

The genus *Albizia* comprises about 150 species widely distributed in the tropics, with the greatest diversity in Africa and South America. *Albizia coriaria* is a tree widely

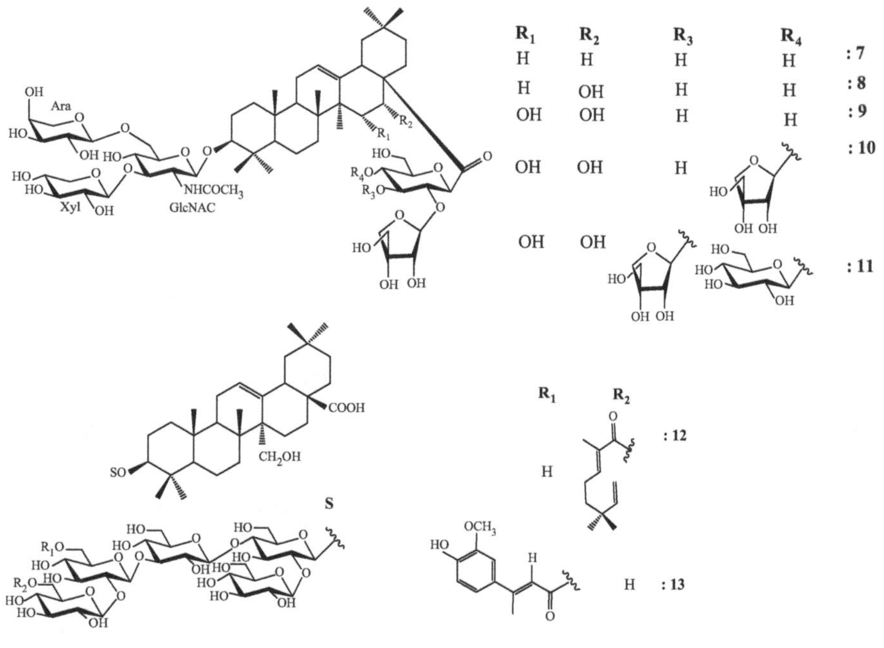

Ara
Xyl
GlcNAC
NHCOCH3
R1 R2 R3 R4
H H H H :7
H OH H H :8
OH OH H H :9
OH OH H :10
OH OH :11
COOH
CH2OH
S
SO
R1 R2
:12
H
OCH3
H :13

FIGURE 3.2 Saponins from *Entada pursaetha* and *Tetrapleura tetraptera*.

distributed in Cameroon and Uganda, where it is harvested for timber or herbal medicine. The bark or root is used as a general tonic and for the treatment of syphilis, skin diseases, jaundice, eye diseases, cough, and sore throat. It is also used to concentrate breast milk in humans [28]. A maceration of the leaves of *A. coriaria* is administered to induce abortion. Leaf decoctions are used externally to treat headache, and as a wash or steam inhalation against fever and toothache. Two new oleanane-type saponins, 3-*O*-{β-D-fucopyranosyl-(1→6)-[β-D-glucopyranosyl-(1→2)]-β-D-glucopyranosyl}-21-*O*-{(2*E*, 6S)-6-*O*-{4-*O*-[(2*E*, 6*S*)-2,6-dimethyl-6-*O*-(β-D-quinovopyranosyl)octa-2,7-dienoyl]-4-*O*-[(2*E*, 6*S*)-2,6-dimethyl-6-*O*-(β-D-quinovopyranosyl)octa-2,7-dienoyl]-β-D-quinovopyranosyl}-2,6-dimethylocta-2,7-dienoyl}acacic acid 28-*O*-α-L-arabinofuranosyl-(1→4)-[β-D-glucopyranosyl-(1→3)]-α-L-rhamnopyranosyl-(1→2)-β-D-glucopyranosyl ester (coriariosides A) (**14**) and 3-*O*-{β-D-fucopyranosyl-(1→6)-[β-D-glucopyranosyl-(1→2)]-β-D-glucopyranosyl}-21-*O*-{(2*E*, 6*S*)-6-*O*-{4-*O*-[(2*E*, 6*S*)-2,6-dimethyl-6-*O*-(β-D-quinovopyranosyl)octa-2,7-dienoyl]-4-*O*-[(2*E*, 6*S*)-2,6-dimethyl-6-*O*-(β-D-quinovopyranosyl)octa-2,7-dienoyl]}-2,6-dimethyl-6-hydroxyocta-2,7-dienoyl}acacic acid 28-*O*-β-D-xylopyranosyl-(1→4)-α-L-rhamnopyranosyl-(1→2)-β-D-glucopyranosyl ester (coriariosides B) (**15**) along with the known 3-*O*-{β-D-fucopyranosyl-(1→6)-[β-D-glucopyranosyl-(1→2)]-β-D-glucopyranosyl}-21-*O*-{(2*E*, 6*S*)-6-*O*-{4-*O*-[(2*E*, 6*S*)-2,6-dimethyl-6-*O*-(β-D-quinovopyranosyl)octa-2,7-dienoyl]-4-*O*-[(2*E*, 6*S*)-2,6-dimethyl-6-*O*-(β-D-quinovopyranosyl)octa-2,7-dienoyl]-β-D-quinovopyranosyl}-2,6-dimethylocta-2,7-dienoyl}acacic acid 28-*O*-β-D-xylopyranosyl-(1→3)-α-L-rhamnopyranosyl-(1→2)-β-D-glucopyranosyl ester (gummiferaoside C) (**16**) were isolated from the roots of *Albizia coriaria* [29].

Furthermore, three new acacic acid derivatives, named coriariosides C, D and E (**17–19**) were also isolated from the roots of *Albizia coriaria*. Their structures were elucidated as 3-*O*-[β-D-xylopyranosyl-(1→2)-β-D-fucopyranosyl-(1→6)-2-(acetamido)-2-deoxy-β-D-glucopyranosyl]-21-*O*-{(2*E*, 6*S*)-6-*O*-{4-*O*-[(2*E*, 6*S*)-2,6-dimethyl-6-*O*-(β-D-quinovopyranosyl) octa-2,7-dienoyl]-4-*O*-[(2*E*, 6*S*)-2,6-dimethyl-6-*O*-(β-D-quinovopyranosyl)octa-2-,7-dienoyl]-β-D-quinovopyranosyl}-2,6-dimethylocta-2,7-dienoyl}acacicacid28-*O*-β-D-xylopyranosyl-(1→4)-α-L-rhamnopyranosyl-(1→2)-β-D-glucopyranosyl ester (**17**), 3-*O*-{β-D-fucopyranosyl-(1→6)-[β -D-glucopyranosyl-(1→2)]-β-D-glucopyranosyl}-21-*O*-{(2*E*, 6*S*)-6-*O*-{4-*O*-[(2*E*, 6*S*)-2,6-dimethyl-6-*O*-(β-D-quinovopyranosyl) octa-2,7-dienoyl]-4-*O*-[(2*E*,6*S*)-2,6-dimethyl-6-*O*-(β-D-quinovopyranosyl)octa-2,7-dienoyl]-β-D-quinovopyranosyl}-2,6-dimethylocta-2,7-dienoyl}acacic acid 28-*O*-α -L-rha mnopyranosyl-(1→2)-β-D-glucopyranosyl ester (**18**) and 3-*O*-[β-D-fucopyranosyl-(1→6)-β-D-glucopyranosyl]-21-*O*-{(2*E*, 6*S*)-6-*O*-{4-*O*-[(2*E*, 6*S*)-2,6-dimethyl-6-*O*-(β-D-quinovopyranosyl)octa-2,7-dienoyl)-β-D-quinovopyranosyl]octa-2,7-dienoyl} acacic acid 28-*O*-β-D-glucopyranosyl ester (**19**) [30] (Figure 3.3).

Entada rheedii grows on every continent adjacent to the Indian Ocean, and in almost every country therein, many different tribes have adapted different uses for this plant. Some tribes believe that the seeds possess magical abilities to bring the owner good luck: the seeds are used as jewelry [31]. In South Africa, the seeds are used as a substitute for coffee. The characteristic feature of the saponins from the genus *Entada* is that glucopyranose, the aglycone-linked sugar in all of them, is N-acylated. Two triterpenoid saponins have been isolated from the seed kernels of *Entada rheedii*, and their structures have been established as; 3-*O*-β-D-xylopyranosyl-(1→3)-*O*-α-L-arabinopyranosyl-(1→6)-2-acetylamino-2-deoxy-D-glucopyranosylentagenic acid 28-*O*-β-apiofuranosyl-(1→3)-β-D-xylopyranosyl-(1→2)-β-D-glucopyranoside (rheediinoside A) (**20**) and 3-*O*-β-D-glucopyranosyl-(1→3)-*O*-[β-D-xylopyranosyl-(1→3)-α-L-arabinopyranosyl-(1→6)]-2-acetylamino-2-deoxy-β-D-glucopyranosylentagenic acid   28-*O*-β-apiofuranosyl-(1→3)-β-D-xylopyranosyl-(1→2)-β-D-glucopyranoside (rheediinoside B) (**21**) [32].

Cylicodiscus gabunensis is a large tree with a cylindrical trunk, reaching 60 m height.

Its bark is used in Cameroon to prepare remedies for headache, filariasis, rheumatism, and gastrointestinal disorders [33]. Gabunoside (**22**), a new oleanane-type triterpene saponin elucidated as 3-*O*-[β-L-arabinopyranosyl-(1→2)-α-L-arabinopyranosyl(1→3)-β-D-glucopyranosyl] maslinic acid-28-α-L-rhamnopy ranosylester, together with the known cylicodiscoside (**23**) and 3-*O*-[α-L-arabinopyranosyl-(1→3)-β-D-glucopyranosyl] cylicodiscic acid (**24**) were isolated from the stem bark of *C. gabunensis* [34] (Figure 3.4).

Albizia lebbeck is a pantropical species distributed in Africa, Asia, America, and Australia. In West Africa, it is traditionally used against diarrhea, dysentery, hemorrhoids, bronchitis, asthma, eczema, and leprosy. In South-East Asia and Australia, the stem bark is used as a folk remedy to treat abdominal tumors, boils, cough, eye disorders, and lung ailments. It is also reported to be astringent, pectoral, rejuvenating, and used as a tonic [35]. Chemical survey of the roots of *A. lebbeck* led to the isolation of two new oleanane-type saponins, named lebbeckosides A and B (**25 and 26**). Their structures were respectively elucidated as

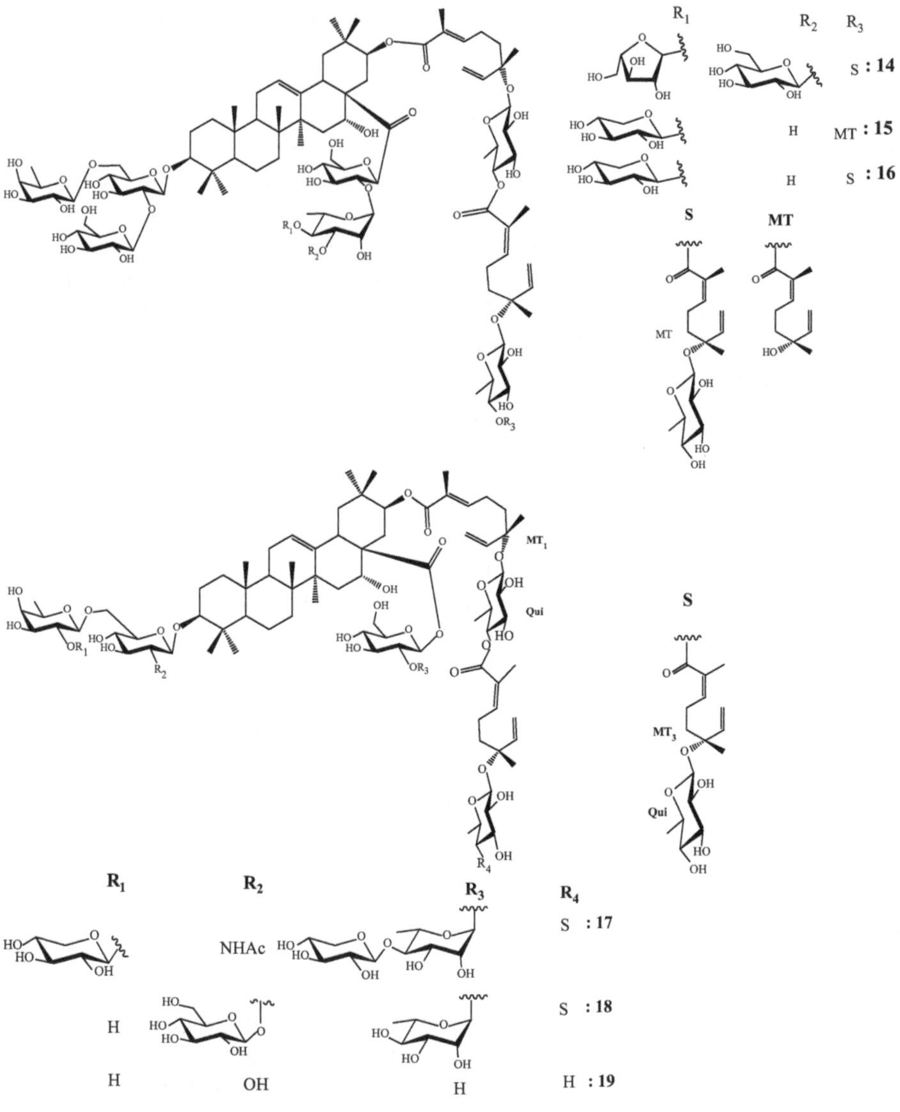

FIGURE 3.3   Saponins from *Albizia coriaria*.

3-O-[β-D-xylopyranosyl-(1→2)-β-D-fucopyranosyl-(1→6)-β-D-glucopyranosyl]-21-O-{(2E,   6S)-6-O-{4-O-[(2E,   6S)-2,6-dimethyl-6-O-(β-D-quinovopyranosyl)octa-2,7-dienoyl]-4-O-[(2E, 6S)-2,6-dimethyl-6-O-(β-D-quinovopyranosyl)octa-2,7-dienoyl]-β-D-quinovopyranosyl}-2,6-dimethylocta-2,7-dienoyl}acacic acid 28-O-β-D-xylopyranosyl-(1→4)-[β-D-glucopyranosyl-(1→3)]-α-L-rhamnopyranosyl-(1→2)-β-D-glucopyranosyl ester (**25**) and 3-O-[β-D-xylopyranosyl-(1→2)-β-D-fucopyranosyl-(1→6)-β-D-glucopyranosyl]-21-O-{(2E, 6S)-6-O-{4-O-[(2E, 6S)-2,6-dimethyl-6-O-(β-D-quinovopyranosyl)octa-2,7-dienoyl]-β-D-quinovopyranosyl}octa-2,7-dienoyl}

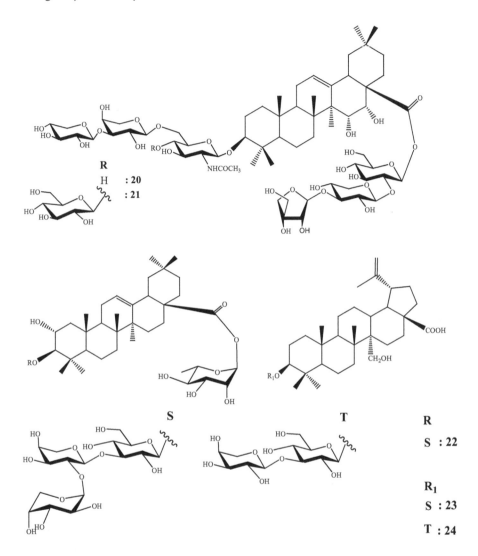

FIGURE 3.4 Saponins from *Entada rhedii* and *Cylicodiscus gabunensis*.

acacic acid 28-*O*-β-D-xylopyranosyl-(1→4)-[β-D-glucopyranosyl-(1→3)]-α-L-rhamnopyranosyl-(1→2)-β-D-glucopyranosyl ester (**26**) [36].

Albizia boromoensis is a large emergent tree of the legume family. It is found growing naturally in Senegal, Sudan, and Cameroon where it is harvested for timber [35], no ethno-medicinal use of this plant is documented. Investigation of the roots of *A. boromoensis* afforded four new oleanane-type saponins, named boromoenosides A-D (**27–30**). The structures of boromoenosides A-D were assigned as; 3-*O*-β-[α-L-arabinopyranosyl-(1→6)-[β-D-xylopyranosyl-(1→2)]-β-D-glucopyranosyl]-28-*O*-[β-D-glucopyranosyl-(1→6)-[β-D-glucopyranosyl-(1→2)]-β-D-glucopyranosyl]-oleanolic acid (**27**), 3-*O*-β-[α-L-

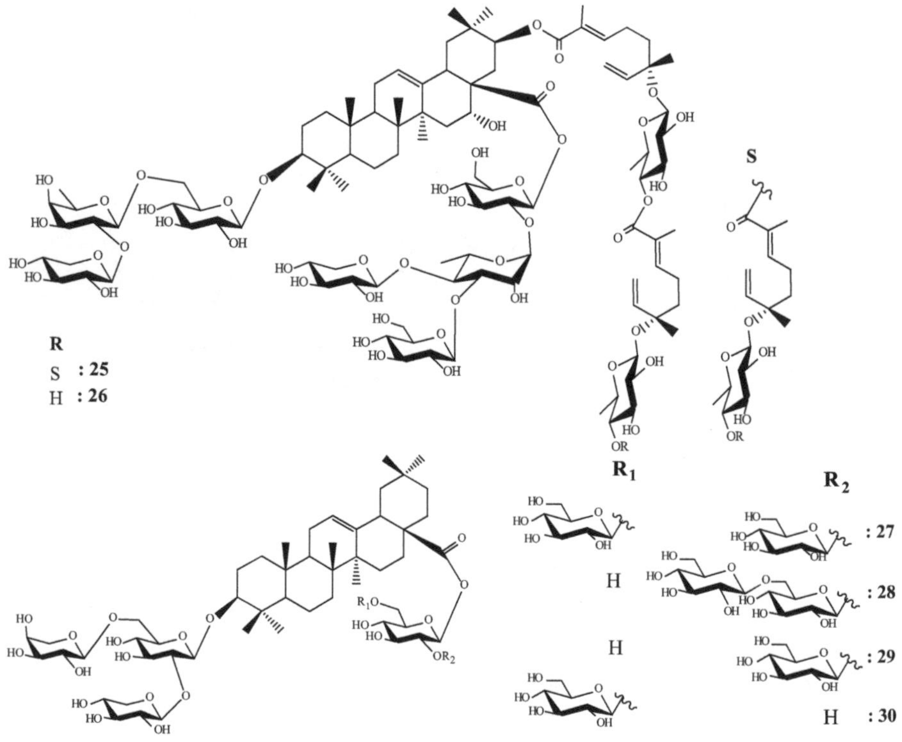

FIGURE 3.5   Saponins from *Albizia lebbeck* and *Albizia boromensis*.

arabinopyranosyl-(1→6)-[β-D-xylopyranosyl-(1→2)]-β-D-glucopyranosyl]
-28-O-[β-D-glucopyranosyl-(1→6)-β-D-glucopyranosyl-(1→2)-β-D-
glucopyranosyl]-oleanolic acid (**28**), 3-O-β-[α-L-arabinopyranosyl-(1→6)-[β-D-
xylopyranosyl-(1→2)]-β-D-glucopyranosyl]-28-O-[β-D-glucopyranosyl-(1→2)-β-D-
glucopyranosyl]-oleanolic acid (**29**), and 3-O-β-[α-L-arabinopyranosyl-(1→6)-[β-D-
xylopyranosyl-(1→2)]-β-D-glucopyranosyl]-28-O-[β-D-glucopyranosyl-(1→6)-β-D-
glucopyranosyl]-oleanolic acid (**30**) [37] (Figure 3.5).

Piptadeniastrum africanum, a large tree up to 50m high widely distributed in
the West and Central Africa [38] is used in Cameroon for the treatment of constipation [39], anemia, meningitis, convulsion, and wound healing [40]. It is also
reported to induce abortion by its oxytocic effects [41]. One new triterpenoid saponin, named piptadeniaoside (**31**), along with two known congeners (**32** and **33**) has
been isolated from the stem bark of *P. africanum*. Their structures were established as
3-O-β-[β-D-xylopyranosyl-(1→2)-α-L-arabinopyranosyl-(1→6)-[α-L-arabinopyranosyl-
(1→3)]-2-acetamido-2-deoxy-β-D-glucopyrano-syl]-28-O-[β-D-apiofuranosyl-(1→3)-β-D-
xylopyranosyl-(1→2)-[β-D-a1piofuranosyl-(1→4)]-β-D-glucopyranosyl]-oleanolic acid (**31**),
3-O-β-[α-L-arabinopyranosyl-(1→2)-α-L-arabinopyranosyl-(1→3)-β-D-glucopyranosyl]
maslinic acid-28-[β-D-glucopyranosyl-(1→2)-α-L-rhamnopyranosyl] ester (**32**) and 3-O-β-

[α-L-arabinopyranosyl-(1→2)-α-L-arabinopyranosyl-(1→3)-β-D-glucopyranosyl]maslinic acid-28-[β-D-glucopyranosyl-(1→6)-β-D-glucopyranosyl-(1→2)-α-L-rhamnopyranosyl] ester (**33**) [42].

Albizia glaberrima bark is applied externally to treat fever, while in Tanzania, a cold-water extract of its root bark is used to treat bilharzia. In Cameroon, twig bark is used dried, and pulverized for the treatment of blenorrhagia and as a decoction to cure chest pain. The ash of its roots is used for the management of liver complaints. Preparations of the leaf and the roots of *A. glaberrima* are also used traditionally in the treatment of epilepsy, anemia, and liver complaints [43].

Phytochemical investigation of the roots of *A. glaberrima* led to the isolation of three new oleanane-type saponins, named glaberrimosides A-C (**34–36**). Their structures were established as 3-*O*-[α-L-arabinopyranosy l-(1→6)-[β-D-glucopyranosyl-(1→2)]-β-D-glucopyranosyl]-28-*O*-[β-D-glucopranosyl-(1→6)-[β-D-glucopyranosyl-(1→2)]-β-D-glucopyranosyl]-oleanolic acid (**34**), 3-*O*-[α-L-arabinopyranosyl-(1→6)-[β-D-glucopyranosyl-(1→2)]-β-D-glucopyranosyl]-28-*O*-[β-D-glucopyranosyl-(1→6)-β-D-glucopyranosyl]-oleanolic acid (**35**), and 3-*O*-[β-D-glucopyranosyl-(1→2)-β-D-glucopyranosyl]-28-*O*-[β-D-glucopyranosyl-(1→6)-[β-D-fucopyranosyl-(1→2)]-β-D-glucopyranosyl]-oleanolic acid (**36**) [44].

Albizia zygia is a medium-sized tree up to 25 m high distributed in Central and Southern Africa, occurring from Senegal to Cameroon and in Sudan [35]. In traditional medicine, its roots barks are used against cough, while its stem bark is used as a purgative and disinfectant. Its leaves are used against diarrhea [35]. In the South Western and Western regions of Cameroon, it is used for the treatment of several diseases such as fever and eczema [45].

Phytochemical investigation of the roots of *Albizia zygia* led to the isolation of two new oleanane-type saponins, named zygiaosides A and B (**37** and **38**) [46] (Figure 3.6).

Acacia albida is a tree growing up to 24 m in height, with a large straight bole, up to 3 m in girth, and rounded crown. In folk medicine, a decoction of the bark of *A. albida* is used in cleansing fresh wounds, similar to that of potassium permanganate, and used as an emetic in fevers by the Masai people of East Africa. It is taken for diarrhea in Tanzania, and for colds, hemorrhage, leprosy, and ophthalmic in West Africa [47,48].

Seven previously undescribed bidesmosidic triterpenoid saponins named albidosides A-G were isolated from a methanol extract of the roots of *A. albida*. Their structures were unambiguously established as 3-*O*-β-D-2-deoxy-2-acetamidoglucopyranosyloleanolic acid-28-*O*-β-D-glucopyr-anosyl-(1→3)-[β-D-xylopyranosyl-(1→4)]-α-L-rhamnopyranosyl-(1→2)-β-D-xylopyranosyl ester (albidoside A) (**39**), 3-*O*-β-D-glucopyranosyl-(1→2)-β-D-glucopyranosyloleanolic acid-28-*O*-β-D-fucopyranosyl-(1→3)-[β-D-xylopyranosyl-(1→4)]-α-L-rhamnopyranosyl-(1→2)-β-D-glucopyranosyl ester (albidosides B) (**40**), 3-*O*-β-D-glucopyranosyl-(1→2)-β-D-glucopyranosyl oleanolic acid-28-*O*-β-D-glucopyranosyl-(1→3)-[β-D-xylopyranosyl-(1→4)]-α-L-rhamnopyranosyl-(1→2)-β-D-xylopyranosyl ester (albidoside C) (**41**), 3-*O*-β-D-glucopyranosyl-(1→4)-β-D-2-deoxy-2-acetamido

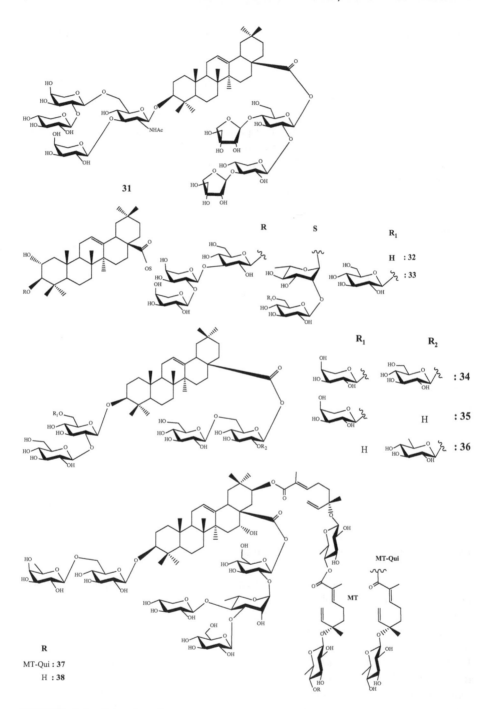

FIGURE 3.6 Saponins from *Piptadeniastrum africanum*, *Albizia glaberrima* and *Albizia zygia*.

glucopyranosyloleanolic acid-28-O-β-D-glucopyranosyl-(1→3)-[β-D-xylopyranosyl-(1→4)]-α-L-rhamno-pyranosyl-(1→2)-β-D-xylopyranosyl ester (albidoside D) (**42**), 3-O-β-D-2-deoxy-2-acetamidoglucopyranosylechinocysticacid-28-O-β-D-fucopyranosyl-(1→3)-[β-D-xylopyranosyl-(→)]-α-L-rhamnopyranosyl-(1→2)-β-D-glucopyranosyl ester (albidoside E) (**43**), 3-O-β-D-glucopyranosyl-(1→4)-β-D-2-deoxy-2-acetamidoglucopyr anosylechinocystic acid-28-O-β-D-glucopyranosyl-(1→3)-[β-D-xylopyranosyl-(1→4)]-α-L-rhamnopyranosyl-(1→2)-β-D-xylopyranosyl ester (albidoside F) (**44**) and 3-O-β-L-arabinopyranosyl-(1→4)-α-L-arabinopyranosyl-(1→4)-β-D-2-deoxy-2-acetamidogl ucopyranosylechinocystic acid-28-O-β-D-glucopyranosyl-(1→3)-[β-D-xylopyranosyl-(1→4)]-α-L-rhamnopyranosyl-(1→2)-β-D-xylopyranosyl ester (albidoside G) (**45**) [49].

Two new triterpene saponins, albidosides H (**46**) and I (**47**) were also reported from the barks of the same medicinal plant. Their structures were elucidated based on extensive 1D- and 2D-NMR studies and mass spectrometry. Albidoside H and I were elucidated as 3-O-β-D-glucopyranosyl-(1→4)-β-D-2-deoxy-2-acetamidoglucopyranosy lechinocystic acid-28-O-β-D-fucopyranosyl-(1→3)-[β-D-xylopyranosyl-(1→4)]-α-L-rhamnopyranosyl-(1→2)-β-D-glucopyranosyl ester (**46**), 3-O-α-L-arabinopyranosyl-(1→4)-α-L-arabinopyranosyl-(1→4)-β-D-2-deoxy-2-acetamidoglucopyranosyl oleanolicacid-28-O-β-D-glucopyranosyl-(1→3)-[β-D-xylopyranosyl-(1→4)]-α-L-rhamn opyranosyl-(1→2)-β-D-xylopyranosyl ester, respectively (**47**) [50] (Figure 3.7).

3.3.1.3 Saponins from Araliaceae

The family of Araliaceae, also known as the ginseng family, comprises 254 species, including *Cussonia bancoensis*, *Cussonia arborea*, *Polyscias fulva*, and *Schefflera abyssinica* considered as richest saponin containing sources. Araliaceae species include trees, shrubs, lianas, and perennial herbaceous plants. The majority of plants in this family occurs in the tropical zones of both hemispheres (Indonesia, Malaysia, and America) but rarely encountered in a temperate climate; some species are cultivated as house plants.

The genus *Cussonia* comprises 21 species distributed in the forests of sub-Saharan Africa; they are traditionally used in the treatment of various diseases such as eye injuries, paralysis, epilepsy, gastro-intestinal problems, women's infertility, diarrhea, stomach ulcers and syphilis [51]. *Cussonia bancoensis* is a medium-sized tree of 15–25 m in height; it is found mostly in the dense humid forest and is widespread from Ivory Coast to Nigeria. *Cussonia* species are used in African traditional medicine mainly against pain, inflammation, gastro-intestinal problems, malaria, and sexually transmitted diseases. *Cussonia bancoensis* is used in Ghana for the management of pain and treating infectious diseases and is also used for making women's combs. It is an African medicinal plant with interesting molluscicidal properties [1].

A new triterpenoid saponin named bafouoside C (3-O-β-D-glucopyranosyl-(1→4)-[β-D-galactopyranosyl-(1→2)]-β-D-glucuronopyranosyloleanolic acid 28-O-β-D glucopyranosyl ester) (**48**), 3-O-β-D-galactopyranosyl-(1→2)-β-D-glucuron opyranosyloleanolic acid (**49**), a new ursane type saponin Zemoside A (**50**) established as 28-O-[α-L-rhamnopyranosyl-(1→4)-β-D-glucopyranosyl-(1→6)-β-D-glucopyranosyl]-23-hydroxyursolic acid ester [52], together with known;

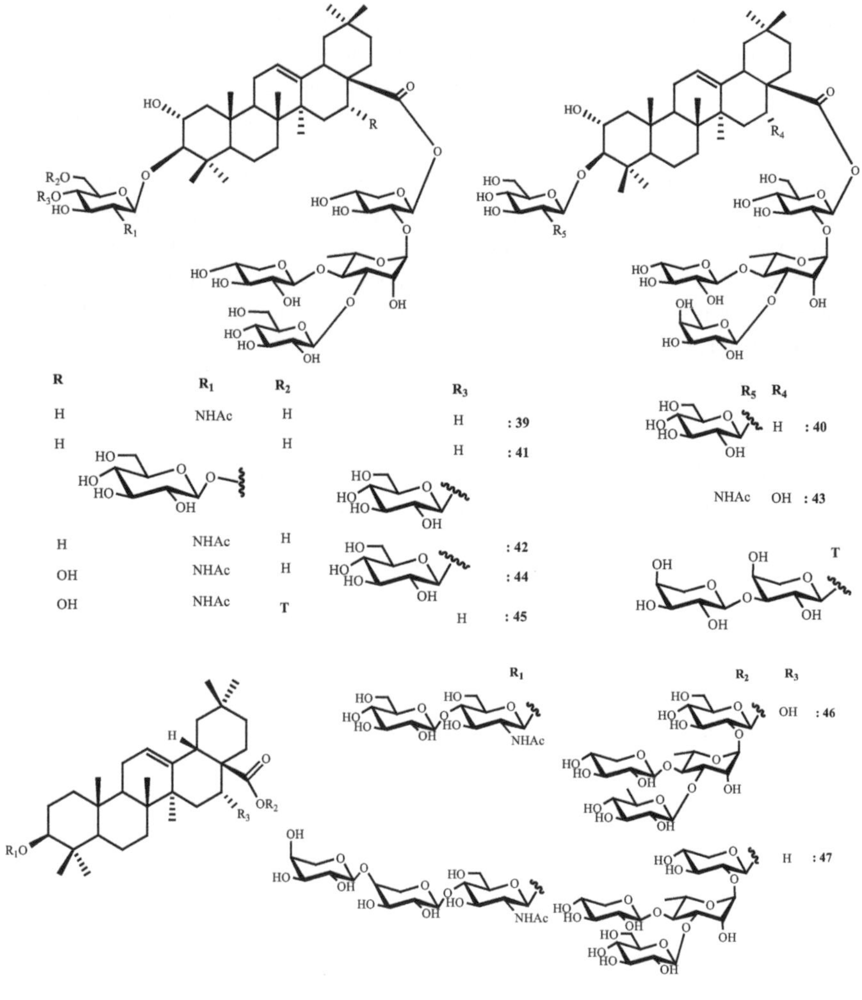

FIGURE 3.7 Saponins *Acacia albida*.

28-O-α-L-rhamnopyranosyl-(1→4)-O-β-D-glucopyranosyl-(1→6)-O-β-D-glucopyranosyl-23 hydroxyursolic acid **(51)**, 3-O-β-D-glucopyranosyl-23-hydroxyursolic acid **(52)** and 3-O-α-L-arabinopyranosyl-23-hydroxyursolic acid **(53)**, were isolated from the roots of *Cussonia bancoensis* [53].

Cussonia arborea is a tree widely distributed in the savanna from western into the central and eastern areas of Africa. Different parts of this plant are used for the treatment of eye injuries, paralysis, epilepsy, convulsions, venereal diseases, diarrhea and women's infertility [38]. Seven saponins, ciwujianoside C3 **(54)**, 3-O-α-L-rhamnopyranosyl-(1→4)-β-D-glucopyranosyl-hederagenin-28-O-β-D-galactopyranosyl-(1→6)-β-D-glucopyranosyl ester (arboreaside A) **(55)**, 3-O-β-D-glucopyranosyl-(1→3)-β-D-xylopyranosyl-(1→2)-α-D-glucopyranosyloleanolic acid 28-O-[α-L-arabinopyranosyl-(1→6)]-[α-L-rhamnopyranosyl-(1→4)]-β-D-glucopy ranosyl ester (arboreaside C) **(56)**, 3-O-[α-L-rhamnopyranosyl-(1→4)]-[β-D-glucopyranosyl-(1→2)]-α-L-arabinopyranosyloleanolic acid 28-O-β-D-galactopyra nosyl-(1→6)-β-D-glucopyranosyl ester (arboreaside D) **(57)**, 3-O-β-D-glucopyranos yloleanolic acid 28-O-α-L-rhamnopyranosyl-(1→2)-β-D-glucopyranosyl-(1→6)-β-D-galactopyranosyl ester (arboreaside E) **(58)**, 23-hydroxyursolic acid 28-O-α-L-rhamnopyranosyl-(1→4)-β-D-glucopyranosyl-(1→6)-β-D-glucopyranosyl ester **(59)**, and 3-O-α-L-rhamnopyranosyl-(1→4)-β-D-glucopyranosyl 23-hydroxyurs-12-en oic acid 28-O-β-D-galactopyranosyl-(1→6)-β-D-glucopyranosyl ester (arboreaside B) **(60)** were isolated from the bark of *Cussonia arborea* [54] (Figure 3.8).

The barks of some *Schefflera* species are used in traditional medicine to treat malaria; extract of their leaves is taken orally to treat rheumatism, whereas the roots are used against inflammations [55]. *Schefflera abyssinica* is planted as an ornamen-tal tree in gardens; the wood is suitable to make internal plates of plywood and makes occasionally for firewood. Eleven triterpenoid glycosides were isolated from the leaves of *S. abyssinica* and their structures were established as: 3-O-α-L-arabinopyranosyl-hederagenin (cauloside A) **(61)**, 3-O-β-D-glucuronopyranosyl-hederagenin (coptero-side B) **(62)**, 3-O-β-D-glucopyranosyl-(1→2)-α-L-arabinopyranosyl-oleanolic acid (fatsiaside C1) **(63)**, 3-O-β-D-glucopyranosyl-(1→3)-α-L-arabinopyranosyl-oleanolic acid (guaianin N) **(64)**, 3- O-β-D-glucopyranosyl-(1→3)-α-L-arabinopyranosyl-hederagenin (collinsonidin) **(65)**, 3- O-β -L-arabinopyranosyl-oleanolic acid-28-O-α-L-rhamnopyranosyl-(1→4)-β-D-glucopyranosyl-(1→6)-β-D-glucopyranosyl ester (ciwujianoside C3) **(54)**, 3-O-α-L-arabinopyranosyl-hederagenin-28-O-α-L-rhamnopyranosyl-(1→4)-β-D-glucopyranosyl-(1→6)-β-D-glucopyranosyl   ester (cauloside D) **(66)**, 3-O-β-D-glucopyranosyl-(1→2)-β-D-glucuronopyranosyl-oleanolic acid-28-O-β-D-glucopyranosyl ester (chikusetsusaponin) **(67)**, 3-O-β-D-glucopyranosyl-(1→2)-α-L-arabinopyranosyl-oleanolic   acid-28-O-α-L-rhamnopyr anosyl-(1→4)-β-D-glucopyranosyl-(1→6)-β-D-glucopyranosyl ester (ciwujianoside A1) **(68)**, 3-O-β-D-glucopyranosyl-(1→3)-α-L-arabinopyranosyl-hederagenin-28-O-α-L-rhamnopyranosyl-(1→4)-β-D-glucopyranosyl-(1→6)-β-D-glucopyranosyl ester **(69)** and 3-O-β-D-glucopyranosyl-(1→2)-α-L-arabinopyranosyl-hederagenin-28-O-α-L-rhamnopyranosyl-(1→4)-β-D-glucopyranosyl-(1→6)-β-D-glucopyranosyl ester (cauloside G) **(70)** [56].

The wood of *Polyscias fulva* is used for interior joinery, doors, musical instru-ments, and containers. In Cameroon it is valued for carving to make handicrafts and masks. In traditional medicine, bark infusions or decoctions are taken for the treat-ment of fever and malaria, as an enema and to treat colic [57]. A bark maceration is applied as drops to the nostrils to treat mental illness. The bark is used in mixtures with other plants to treat epilepsy. Leaf decoctions are taken to treat intestinal com-plaints and dermatosis caused by parasites [58]. The flowers are a good source of nec-tar and pollen for honeybees. Three triterpene glycosides including 12-oxo-3β, 16β, 20(S)-trihydroxydammar-24-ene-3-O-α-L-rhamnopyranosyl-(1→2)-β-D-glucopy ranoside (polysciasoside A) **(71)**, α-hederin **(72)**, 3-O-α-L-rhamnopyranosyl-(1→2)-α-L-arabinopyranosyl-hederagenin-28-O-α-L-rhamnopyranosyl-(1→4)-β-D-glucopyranosyl-(1→6)-β-D-glucopyranosyl ester **(73)**, have been previously isolated from the leaves of *Polyscias fulva* [59]. Later, eight other triterpene glycosides have also been isolated from the stem bark of this plant and their structures were

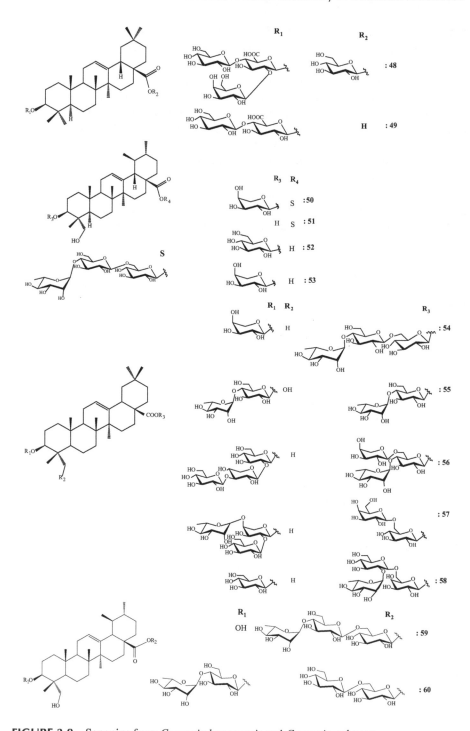

FIGURE 3.8   Saponins from *Cussonia bancoensis* and *Cussonia arborea*.

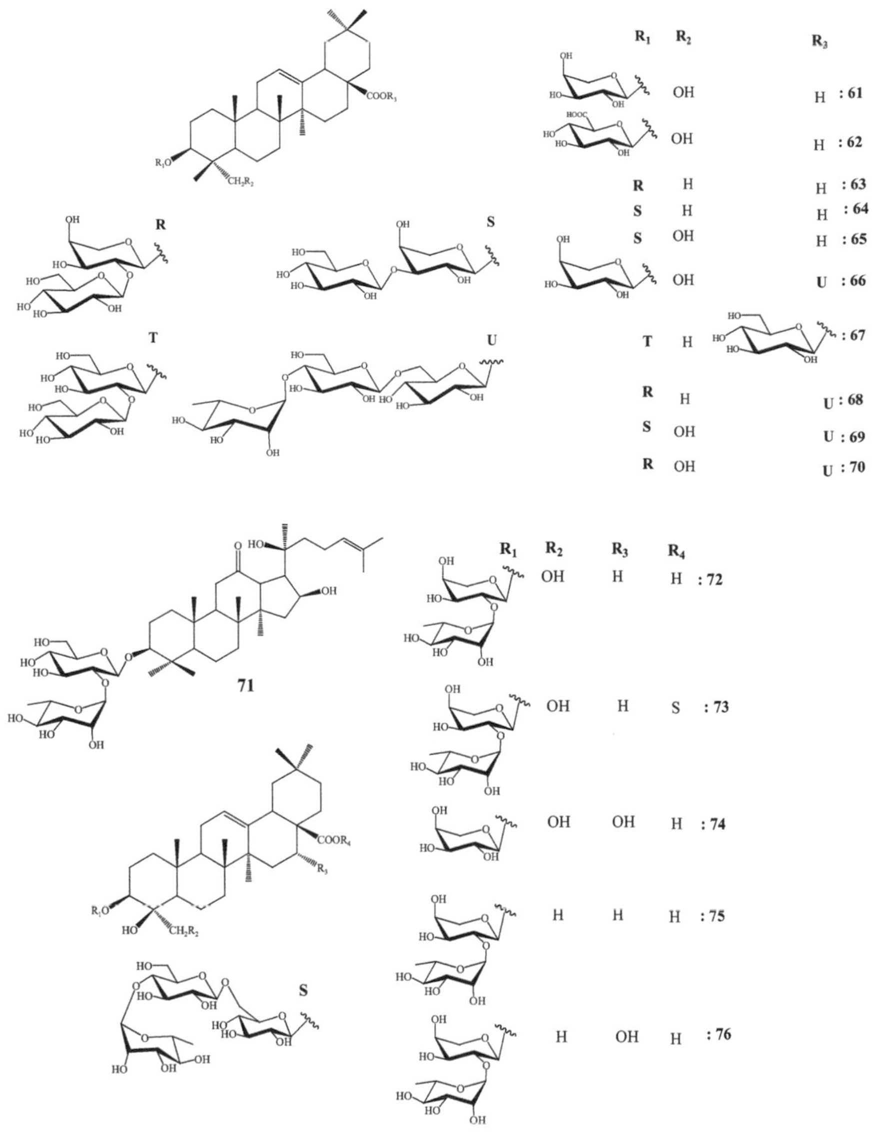

FIGURE 3.9 Saponins from *Schefflera abysinica* and *Polyscia fulva*.

elucidate as: α-hederin (**72**), 3-*O*-α-L-rhamnopyranosyl-(1→2)-α-L-arabinopyranosyl-hederagenin-28-*O*-α-L-rhamnopyranosyl-(1→4)-β-D-glucopyranosyl-(1→6)-β-D-glucopyranosyl ester (**73**), 3-*O*-α-L-arabinopyranosyl-hederagenin (cauloside A) (**61**), 3-*O*-α-L-arabinopyranosyl-collinsogenin (collinsonin) (**74**), 3-*O*-[α-L-rhamnopyranosyl-(1→2)-α-L-arabinopyranosyl]-oleanolic acid (β-hederin) (**75**) and 3-*O*-[α-L-rhamnopyranosyl-(1→2)-α-L-arabinopyranosyl]-echinocystic acid (**76**) [60] (Figure 3.9).

3.3.1.4 Saponins from Combretaceae

The Combretaceae are trees, shrubs, and lianas comprising about 20 genera and 600 species. Widespread in the subtropics and tropics, it represents a large family of plants; the most commonly occurring genera are *Combretum* and *Terminalia*. Some members of this family produce useful construction timber, such as idigbo from *Terminalia ivorensis* [61]. Most Africans use boiled root decoction of *Combretum molle* to treat constipation, headaches, stomachs aches, fever, dysentery, and swellings. The leaves are chewed and soaked in water and the juice drunk for chest complaints [62].

The MeOH extract of the stem bark of *Combretum molle* was evaluated for anti-inflammatory activity [63], followed by fractioning with EtOAc and n-BuOH and both fractions showed anti-inflammatory activity, thereafter their purification afforded three saponins including a new oleanane-type triterpene saponin, β-D-glucopyranosyl-2α, 3β, 6β-trihydroxy-23-galloylolean-12-en-28-oate (**77**), together with two known oleanane-type pentacyclic triterpenoids, arjunglucoside I (**78**), and combreglucoside (**79**) [63].

Terminalia is a genus of large trees of the flowering plant, comprising around 100 species. This genus gets its name from Latin *terminus*, referring to the fact that the leaves appear at the very tips of the shoots. Decoctions of powder bark of *Terminalia ivorensis* are used in traditional medicine to treat wounds, ulcers, and hemorrhoids, against malaria and yellow fever. Sap of the leaves is applied on cuts and, it is also taken together with bark decoctions to treat gonorrhea and kidney complaints.

Three new saponins including two dimeric triterpenoid glucosides and a monomeric triterpenoid saponin together with the known sericoside 2α, 3β, 19α, 24-tetrahydroxyolean-12-en-28-oic acid (**80**), were isolated from the bark of *Terminalia ivorensis*. The structures of these new metabolites were established as a dimer of β-D-glucopyranosyl-18,19-seco-2α, 3α, 19,19,24-pentahydroxyolean-12-en-28-oate and β-D-glucopyranosyl-2α, 3β, 19α, 24-tetrahydroxyolean-12-en-28-oate (ivorenoside A) (**81**), a dimer of β-D-glucopyranosyl-18,19-seco-24-carboxyl-2α, 3β, 19,19-tetrahydroxyolean-12-en-28-oate and β-D-glucopyranosyl-2α, 3β, 19α, 24-tetrahydroxyolean-12-en-28-oate (ivorenoside B) (**82**) and β-D-glucopyranosyl-2α, 3β, 19β, 24-tetrahydroxyolean-11-oxo-olean-12-en-28-oate (ivorenoside C) (**83**) [64]. The structure elucidation of these dimeric compounds required an extensive use of tandem MS-MS analysis (Figure 3.10).

3.3.1.5 Saponins from Sapotaceae

The Sapotaceae, a family of flowering plants, includes about 800 species of evergreen trees and shrubs distributed in about 65 genera and includes the genus *Butyrospermum*. *Butyrospermum parkii* is a small tree that grows up to 14 m high. The seeds of this plant are used in traditional medicine to treat scabies, ulcers, wounds, and nasal stiffness [65,66]. Three new triterpenoid saponins, elucidated as; 3-*O*-β-D-glucopyranosyloleanolic acid 28-*O*-β-D-xylopyranosyl-(1→4)-α-L-rhamnopyranosyl-(1→2)-β-D-xylopyranoside (parkioside A) (**84**), 3-*O*-[β-D-apifuranosyl-(1→3)- β-D-glucopyranosyl]oleanolic acid 28-O-[β-D-apifuranosyl-(1→3)-β-D-xylopyranosyl-(1→4)-[α-L-rhamnopyranosyl-(1→3)]-α-L

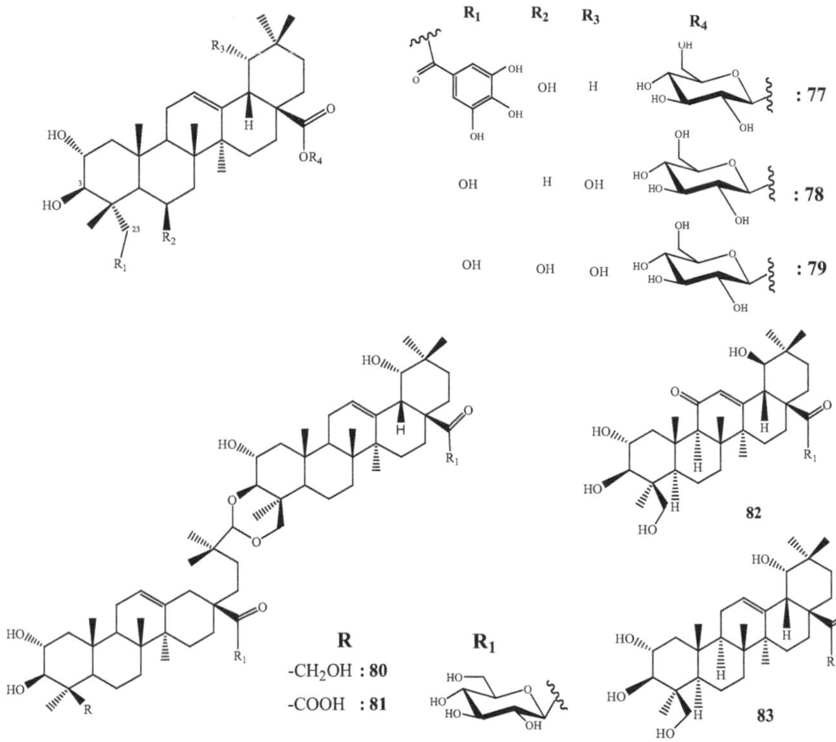

FIGURE 3.10 Saponins from *Combretum molle* and *Terminalia ivorensis*.

rhamnopyranosyl-(1→2)-β-D-xylopyranoside (parkioside B) (**85**) and 3-*O*-β-D-glucuronopyranosyl-16-α-hydroxyprotobassic acid 28-*O*-α-L-rhamnopyranosyl-(1→3)-β-D-xylopyranosyl-(1→4)-α-L-rhamnopyranosyl-(1→2)-β-Dxylopyranoside (parkioside C) (**86**) and the know androseptoside (3-*O*-β-D-glucopyranosyloleanolic acid) (**87**) were isolated from the *n*-BuOH extract of the root bark of *Butyrosper mum parkii* [67].

Chemical studies of the EtOAc extract of *Gambeya boukokoensis* stem bark led to the isolation of two new saponins, elucidated as 2β, 3β, 6β, 28-tetrahydroxyolean-12-en-23-oic acid 23-*O*-α-L-arabinopyranosyl ester (gamboukokoenside A) (**88**) and 6β, 28-dihydroxy-3-oxoolean-12-en-23-oic acid 23-*O*-α-L-arabinopyranosyl ester (gamboukokoenside B) (**89**) [68] (Figure 3.11).

3.3.1.6 Saponins from Melianthaceae

All members of Melianthaceae are trees or shrubs, flowering plants found in tropical and southern Africa. Melianthaceae or the honey bush family consists of 3 genera (*Melianthus*, *Bersama*, and *Greyia*) and 11 species [69]. *Bersama* species contain shrubs to small trees with leaves serrate leaflet edges. The genus *Bersama* comprises four species (*B. swinnyi*, *B. yangambiensis*, *B. abyssinica*, and *B. engleriana*) all of which are very tall trees; the last two species are found in Cameroon. Five new

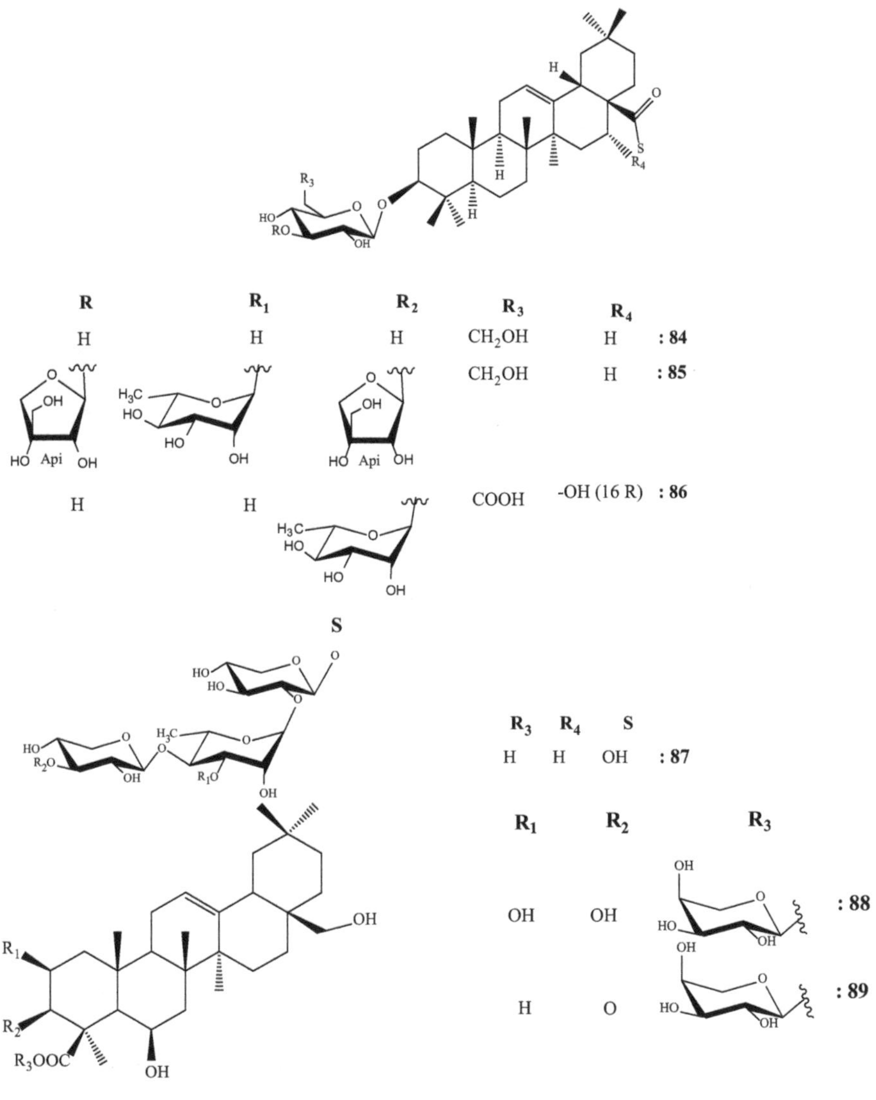

FIGURE 3.11   Saponins from *Butyrospermun parkii* and *Gambeya boukokoensis.*

3-*O*-glucuronide triterpenes (**90–94**) were isolated from the stem bark of *Bersama engleriana*. Their structures were established as 3-*O*-[β-D-glucopyranosyl-(1→2)-β-D-glucuronopyranosyl]-28-*O*-[β-D-glucopyranosyl]-betulinic acid (**90**), 3-*O*-[β-D-glucopyranosyl-(1→2)-[β-D-galactopyranosyl-(1→3)]-β-D-glucuronopyranosyl]-oleanolic acid (**91**), 3-*O*-[β-D-glucopyranosyl-(1→3)-β-D-glucuronopyranosyl]-28-*O*-[β-D-xylopyranosyl-(1→6)-β-D-glucopyranosyl]-oleanolic acid (**92**), 3-*O*-[β-D-galactopyranosyl-(1→3)-β-D-glucuronopyranosyl]-28-*O*-[β-D-glucopyranosyl-(1→4)-β-D-glucopyranosyl]-oleanolic acid (**93**) and 3-*O*-[β-D-glucopyranosyl-(1→3)-β-D-galactopyranosyl-(1→3)-β-D-glucuronopyranosyl]-28-*O*-[β-D-xylopyranosyl-(1→6)-β-D-glucopyranosyl]-oleanolic acid (**94**) [70].

3.3.1.7 Saponins from Onagraceae

Onagraceae are mostly herbaceous although some imported from South and Central American species become woody. They can be aquatic or grow in mud, rocks, or grassy plains [71]. *Ludwigia leptocarpa* is used in Nigerian folk medicine for the treatment of rheumatism and dysentery [38]. Purification of the *n*-butanol soluble fraction of the crude MeOH extract of whole plant of *Ludwigia leptocarpa* afforded four new saponins, leptocarposide A-D (**95–98**) [72,73].

3.3.1.8 Saponins from Apiaceae

Apiaceae synonym of Umbelliferae, is a family of mostly aromatic and some wonderful edible plants commonly found in the celery, carrot, or parsley families. It is the largest family of plants, with more than 3700 species comprising between 300 and 400 genera, including *Hydrocotyle bonariensis*. This plant has been known for treating leprosy and tuberculosis; relieving the pain of rheumatism [74].

Phytochemical investigation of the underground parts of *H. bonariensis* led to the isolation of five new oleanane-type triterpenoid saponins; bonarienosides A-E (**99–103**) elucidated as 3-*O*-β-D-glucopyranosyl-(1→2)-[α-L-arabinopyranosyl-(1→3)]-β-D-glucuronopyranosyl-21-*O*-(2-methylbutyroyl)-22-*O*-acetyl-R$_1$-barrigenol (**99**), 3-*O*-{β-D-glucopyranosyl-(1→2)-[α-L-arabinopyranosyl-(1→3)]-β-D-glucuronopyranosyl}-21-*O*-(2-methylbutyroyl)-28-*O*-acetyl-R$_1$-barrigenol(**100**),3-*O*-{β-D-glucopyranosyl-(1→2)-[α-L-arabinopyranosyl-(1→3)]-β-D-glucuronopyranosyl}-21-*O*-acetyl-R$_1$-barrigenol (**101**), 3-*O*-{β-D-glucopyranosyl-(1→2)-[α-L-arabinopyranosyl-(1→3)]-β-D-glucuronopyranosyl}-R$_1$-barrigenol (**102**) and 3-O-{β-D -glucopyranosyl-(1→2)-[α-L-arabinopyranosyl-(1→3)]-β-D-glucuronopyranosyl}-22-*O*-(2-methylbutyroyl)-A$_1$-barrigenol (**103**), together with saniculoside-R1 (**104**) [75] (Figure 3.12).

3.3.1.9 Saponins from Tiliaceae

The Tiliaceae are trees, shrubs, or rarely herbs comprising about 50 genera and 450 species among which *Triumfetta cordifolia*. This plant is a fast-growing, sparsely branched shrub usually growing up to 5 m. A multi-purpose plant, mainly used supplying food, medicines, fiber and several other commodities. It has sometimes been an important item of commerce, with stems sold for food and the fiber has in the past use for making packaging material. *T. cordifolia*, located in tropical Africa, is a shrub about 5 m high, used in Cameroon as foodstuff [76]. A triterpenoid saponin dimer, named (3β)-stigmasta-5,22-diene-3,29-diol 3-propanoate 29-triacontanoate and (2α, 3β, 19α)-2,3,19-trimethoxyurs-12-ene-24,28-dioic acid 24-[(2α, 3β)-24,28-bis(β-D-glucopyranosyloxy)-2-hydroxy-24,28-dioxours-12-en-3-yl] ester (**105**) was isolated from the leaves of *Triumfetta cordifolia* [76].

3.3.1.10 Saponins from Myrsinaceae

Ardisia kivuensis is used in traditional medicine to treat venereal diseases. Column chromatography of the *n*-butanol extract of the stem of *Ardisia kivuensis* (Myrsinaceae) led to the isolation of a new triterpenoid saponin, ardisikivuoside (**106**) elucidated as 3-*O*-β-D-xylopyranosyl-(1→3)-β-D-glucopyranosyl-(1→4)-β-D-xylopyranosyl-3β-hydroxy-13β, 28-epoxyoleanan-16-oxo-30-al [77] (Figure 3.13).

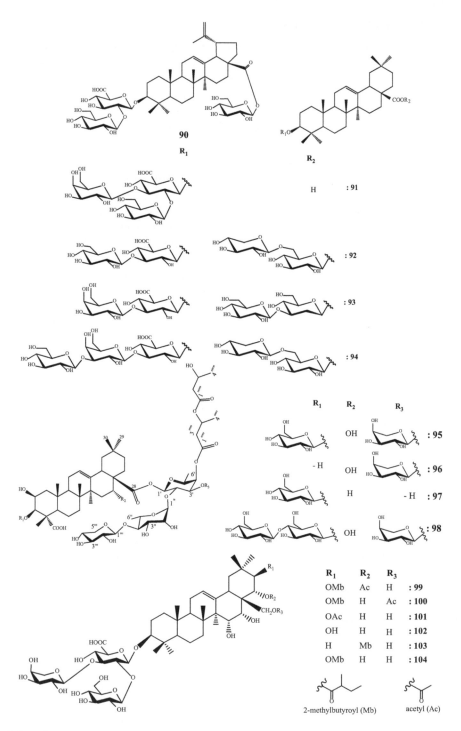

FIGURE 3.12   Saponins from *Bersama engleriana*, *Ludwigia leptocarpa*, and *Hydrocotyle bonariensis*.

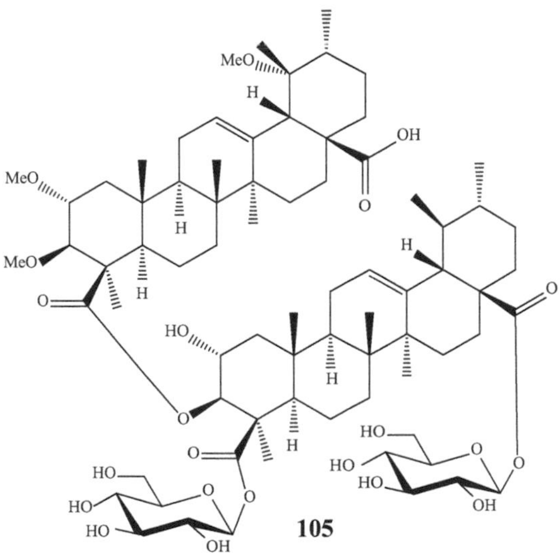

105

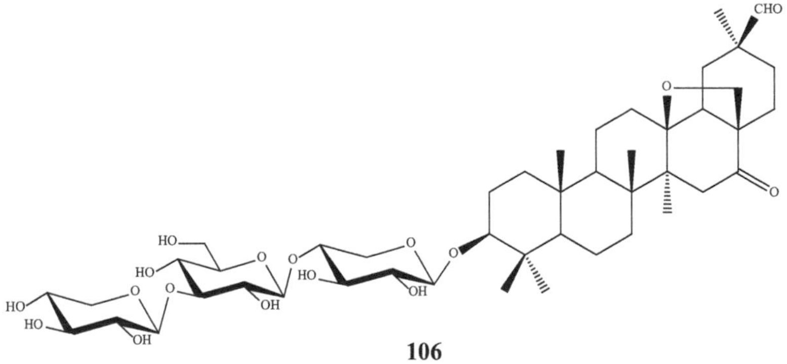

106

FIGURE 3.13 Saponins from *Triumfetta cordifolia* and *Ardisia kivuensis*.

3.3.2 STEROIDAL SAPONINS

3.3.2.1 Saponins from Dracaenaceae

The family of the Dracaenaceae includes about 480 species distributed in the arid tropical and subtropical regions. Dracaenaceae plants from Africa can reach several meters tall. The flowers meet at the end of the stem and are of green-yellow color. Some species of Dracaenaceae, example of *Sansevieria trifasciata* are cultivated as garden or house plants. Many species of the genus *Dracaena* are used in African traditional medicine for the treatment of variety diseases [78].

The genus *Dracaena* includes 40 to 100 species distributed in tropical and subtropical dry climate regions throughout the world. They are used in African traditional medicine for the treatment of a variety of diseases. A *Dracaena mannii* bark decoction is used by the native people of the western highlands of Cameroon against abdominal pains. A new steroidal saponin, mannioside A (**107**), was isolated from the stem bark of *Dracaena mannii*, together with the known pennogenin-3-*O*-β-D-glucopyranoside

(**108**) and pennogenin-3-*O*-α-L-rhamnopyranosyl-(1→2)-[α-L-rhamnopyranosyl-(1→3)]-β-D glucopyranoside (**109**) [79].

Dracaena deisteliana and *Dracaena arborea* are used in African folk medicine for the treatment of heart and kidney problems, measles, eye injury, paralysis, epilepsy, convulsions, spasms, venereal diseases, diarrhea, dizziness, and infertility in women, and for social purposes [80].

Four new steroidal saponins were isolated from the stem and bark of two species of *Dracaena*, deistelianosides A and B (**110** and **111**) from *D. deisteliana* and arborea saponins A and B (**112** and **113**) from *D. arborea*. Their structures were established as: diosgenin 3-*O*-[3-*O*-sulfate-α-L-rhamnopyranosyl-(1→4)]-β-D-glucopyranoside (**110**), 1-*O*-β-D-xylopyranosyl-(1→2)-[α-L-rhamnopyranosyl-(1→3)]-β-D-fucopyran osyl (23S, 24S)-spirosta-5,25(27)-diene-1β, 3β, 23α, 24α-tetrol 24-*O*-α-L-arabinop yranoside (**111**), pennogenin-3-*O*-α-L-rhamnopyranosyl-(1→2)-[α-L-rhamnopyrano syl-(1→3)]-[6-*O*-acetyl]-β-D-glucopyranoside (**112**) and 24R-hydroxypennogenin 3-*O* -α-L-rhamnopyranosyl-(1→2)-[α-L-rhamnopyranosyl-(1→3)]-β-D-glucopyran oside (**113**) [81].

The genus *Sansevieria* comprises 60 species worldwide, found mainly in dry and arid areas of the Old-World tropics and subtropics. Members of *Sansevieria* are of great economic importance as ornamentals, source of fiber and as medicine for curing different ailments. *Sansevieria trifasciata* commonly known as "snake plant" is an evergreen herbaceous perennial plant used for the treatment of inflammatory conditions, boils, and fever. Some species have an ethnopharmacological background, in particular *S. trifasciata* is used to treat victims of snake bite, ear pain, swellings, and fever [82,83].

Four previously unreported steroidal saponins, trifasciatosides A, B, C and D (**114** and **115, 117** and **118**), three pairs of previously undescribed steroidal saponins, trifasciatosides E and F (**120** and **121**), trifasciatosides G and H (**128** and **129**) and trifasciatosides I and J (**126** and **127**), together with twelve known saponins (**116, 119, 122–125, 130–135**) were isolated from the *n*-butanol soluble fraction of the methanol extract of *S. trifasciata* [84].

Phytochemical investigation of the aerial parts of the same plant has also resulted in the isolation of four previously unreported steroidal saponins as two pairs of inseparable regioisomers elucidated as a mixture of (23S)-3β, 23-dihydroxyspirosta-5,25(27)-dien-1β-yl*O*-(2-*O*-acetyl-α-L-rhamnopyranosyl)-(1→2)-*O*-[β-D-xylopyranosyl-(1→3)]-α-L-arabinopyranoside (trifasciatoside K) (**136**) and (23S)-3β, 23-dihydrox yspirosta-5,25(27)-dien-1β-yl *O*-(3-*O*-acetyl-α-L-rhamnopyranosyl)-(1→2)-*O*-[β-D-xyl opyranosyl-(1→3)]-α-L-arabinopyranoside (trifasciatoside L) (**137**) and the mixture of (23S, 24S)-3β, 23,24-trihydroxyspirosta-5,25(27)-dien-1β-yl *O*-(2-*O*-acetyl-α -L-rhamnopyranosyl)-(1→2)-*O*-[β-D-xylopyranosyl-(1→3)]-α-L-arabinopyranoside (trifasciatoside M) (**138**) and (23S, 24S)-3β, 23,24-trihydroxyspirosta-5,25(27)-dien-1β-yl *O*-(3-*O*-acetyl-α-L-rhamnopyranosyl)-(1→2)-*O*-[β-D-xylopyranosyl-(1→3)]-α-L-arabinopyranoside (trifasciatoside N) (**139**) [85] (Figure 3.14).

3.3.2.2 Saponins from Dioscoreaceae

The Dioscoreaceae constitutes a family consisting mainly of tropical climbers. It comprises 350–400 species [86], and it is distributed throughout the tropics and

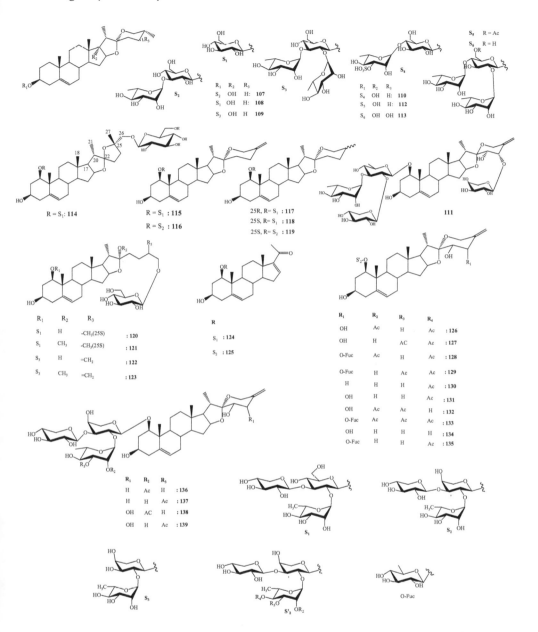

FIGURE 3.14 Saponins from *Dracaena mannii*, *Dracaena deisteliana*, *Dracaena arborea*, and *Sansevieria trifasciata*.

subtropics regions especially in West Africa, parts of Central America and the Caribbean, the Pacific Islands and Southeast Asia.

Dioscorea bulbifera is used by tribal people in Bangladesh for treatment of leprosy and tumors. Its bubils are used by the native people of the western highlands of Cameroon for the treatment of pig Cysticercosis, though the tubers after collection during the farming period are destroyed and burned because of their intense bitterness [87]. Eleven steroidal saponins, dioscoreanosides A-K (**140–150**), together with five known saponins, which were identified by comparison of their spectroscopic data with literature values as pennogenin 3-*O*-α-L-rhamnopyranosyl-(1→4)-α-L-rhamnopyranosyl-(1→4)-[α-L-rhamnopyranosyl-(1→2)]-β-D-glucopyranoside (**151**), 23β, 27-dihydroxy-pennogenin 3-*O*-α-L-rhamnopyranosyl-(1→4)-α-L-rhamnopyranosyl-(1→4)-[α-L-rhamnopyranosyl-(1→2)]-β-D-glucopyranoside(**152**), spiroconazole A (**153**), floribundasaponin B (**154**), and 26-*O*-β-D-glucopyranosyl-(25*R*)-5-en-furost-3β, 17α, 22α, 26-tetraol-3-*O*-α-L-rhamnopyranosyl-(1→4)-α-L-rhamnopyranosyl-(1→4)-[α-L-rhamnopyranosyl-(1→2)]-β-D-glucopyranoside (**155**) were isolated from the flowers of *D. bulbifera* [88].

Three new steroidal saponins, named diospreussinosides A-C (**156–158**), along with two known ones (**148** and **151**) were isolated from rhizomes of *Dioscorea preussii*. Their structures were elucidated as (25*S*)-17α, 25-dihydroxyspirost-5-en-3β-yl-*O*-α-L-rhamnopyranosyl-(1→4)-α-L-rhamnopyranosyl-(1→4)-β-D-glucopyranoside (**156**), (25S)-17α, 25-dihydroxyspirost-5-en-3β-yl-*O*-α-L-rhamnopyranosyl-(1→4)-α-L-rhamnopyranosyl-(1→4)-[α-L-rhamnopyranosyl-(1→2)]-β-D-glucopyranoside (**157**) and (24S, 25R)-17α, 24,25-trihydroxyspirost-5-en-3β-yl-*O*-α-L-rhamnopyranosyl-(1→4)-α-L-rhamnopyranosyl-(1→4)-[α-L-rhamnopyranosyl-(1→2)]-β-D-glucopyranoside (**158**) [89] (Figure 3.15).

3.3.2.3 Saponins from Agavaceae

Agavaceae is a family of plants growing in dry and hot climates, some plants of this family grow in the desert. Well-known plants of this family are the agave, yucca, and Joshua tree [90].

The genus *Furcraea* has about 20 species mainly cultivated as ornamental plants. Nevertheless, leaves of some plants of this genus have been used to treat ulcers, venereal diseases, rheumatism and as diuretic. They are also used as a source of fiber and are used to reduce swelling [91]. *Furcraea bedinghausii* is a large succulent plant from southern Mexico with long blue-green leaves at the top of an 8–12-foot-tall trunk. Phytochemical study of the roots of this plant led to the isolation of two pairs of steroidal saponins: (25*R*)-2α-3β-dihydroxy-5α-spirostan-12-one   3-*O*-β-D-glucopyranosyl-(1→2)-*O*-[β-D-xylopyranosyl-(1→3)]-*O*-β-D-glucopyranosyl-(1→4)-β-D galactopyranoside (**159**) and (25*R*)-2α-3β-dihydroxy-5α-spirost-9-en-12-one 3-*O*-β-D-glucopyranosyl-(1→2)-*O*-[β-D-xylopyranosyl-(1→3)]-*O*-β-D-glucopyranosyl-(1→4)-β-D-galactopyranoside (**160**), (25*R*)-3β-hydroxy-5α-spirostan-12-one 3-*O*-β-D-glucopyranosyl (1→2)-*O*-[β-D-xylopyranosyl-(1→3)]-*O*-β-D-glucopyranosyl-(1→4)-β-D-galactopyranoside (**161**), and (25*R*)-3β-hydroxy-5α-spirost-9-en-12-one   3-*O*-β-D-glucopyranosyl-(1→2)-*O*-[β-D-xylopyranosyl-(1→3)]-*O*-β-D-glucopyranosyl-(1→4)-β-D-galactopyranoside (**162**) [90].

Cordyline fruticosa comprises more than 480 species distributed in tropical and subtropical regions of the World. The plant is widely used in both everyday life and traditional medicine. A hedge of *Cordyline fruticosa* around the house was believed to ward off evil spirits and bring good luck [92]. The leaves have been used for costumes, decorations, clothing, sandals, packaging, and cooking. In addition, the plant is traditionally used for the treatment of various diseases: the leaves are used to treat

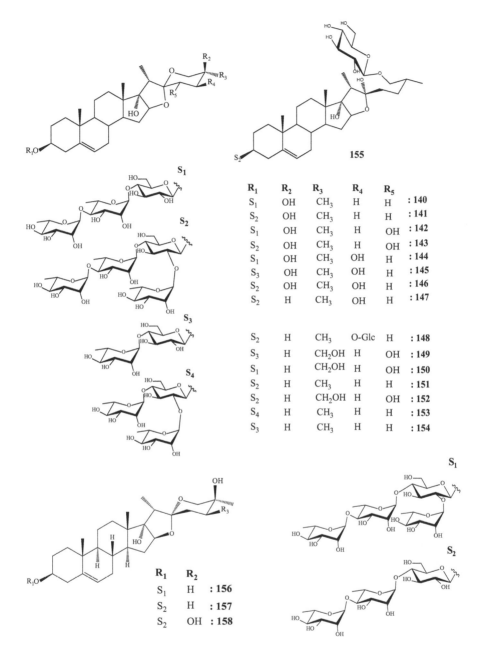

	R_1	R_2	R_3	R_4	R_5	
S_1	OH	CH_3	H	H	: 140
S_2	OH	CH_3	H	H	: 141
S_1	OH	CH_3	H	OH	: 142
S_2	OH	CH_3	H	OH	: 143
S_1	OH	CH_3	OH	H	: 144
S_3	OH	CH_3	OH	H	: 145
S_2	OH	CH_3	OH	H	: 146
S_2	H	CH_3	OH	H	: 147
S_2	H	CH_3	O-Glc	H	: 148
S_3	H	CH_2OH	H	OH	: 149
S_1	H	CH_2OH	H	OH	: 150
S_2	H	CH_3	H	H	: 151
S_2	H	CH_2OH	H	OH	: 152
S_4	H	CH_3	H	H	: 153
S_3	H	CH_3	H	H	: 154

	R_1	R_2	
S_1	H	: 156
S_2	H	: 157
S_2	OH	: 158

FIGURE 3.15 Saponins from *Dioscorea bulbifera* and *Dioscorea preussii*.

sore throat and neck pain, as hemostatic, and also to induce abortion; the roots are used for toothache, laryngitis, and infections of mammary glands [93,94]. Three new steroidal saponins, spirosta-5,25(27)-diene-1β, 3β-diol-1-*O*-α-L-rhamnopyranosyl-(1→2)-β-D-fucopyranoside (fruticoside H) (**163**), 5α-spirost-25(27)-ene-1β, 3β-diol-1-*O*-α-L-rhamnopyranosyl-(1→2)-(4-*O*-sulfo)-β-D-fucopyranoside (fruticoside I)

(**164**) and (22*S*)-cholest-5-ene-1β, 3β, 16β, 22-tetrol-1-*O*-β-galactopyranosyl-16-*O*-α-L-rhamnopyranoside (fruticoside J) (**165**) were reported from the leaves of *Cordyline fruticosa* [95].

Furcraea tuberosa reaches heights of 2–8 m and its leaves are lanceolate with spiny margins, commonly known as Female Karata, is a perennial shrub. Originally, it is a West Indian endemic, and in spite of the fact that it is cultivated in various parts of the world as an ornamental plant it has recently been recorded as having become naturalized in the Mpumalanga province of South Africa [96]. Haitian people use the fibers of this plant to produce hammocks and cordage [97]. Phytochemical study of the air-dried mature fruits of *F. tuberosa* led to isolation of nine steroidal saponins among which one new spirostanol bidesmoside. They were identified as (25*R*)-6α-(β-D-glucopyranosyloxy)-5α-spirostane-3β-*O*-[(6-*O*-hexadecanoyl)-β-D-glucopyranoside] (**166**), (25*R*)-6α-hydroxy-5α-spirostan-3β-yl ß-D-glucopyranoside (chlorogenin-3-*O*-β –D-glucopyranoside (**167**), (25*R*)-5α-spirostane-3β, 6α-diyl bis β-D-glucopyranoside (**168**), (23*S*, 25*R*)-23-hydroxy-5α-spirostane-3β, 6α-diyl bis β-D-gluco-pyranoside (cantalasaponin-1) (**169**), (25*R*)-3β-[(*O*-β-D-glucopyranosyl-(1→2)-*O*-[*O*-α-L-rhamnopyranosyl-(1→4)-β-D-glucopyranosyl-(1→3)]-*O*-β-D-glucopyranosyl-(1→4)-β-D-galacto-pyranosyl)oxy]-5α-spirostan-12-one(**170**),(25*R*)-5α-spirostan-3β–yl-*O*-β-D-glucopyranosyl-(1→2)-*O*-[*O*-α-L-rhamnopyran osyl-(1→4)-β-D-glucopyranosyl-(1→3)]-*O*-β-D-glucopyranosyl-(1→4)-β-D-galactopyranoside (**171**), (25*R*)-3β-[*O*-β-D-glucopyranosyl-(1→3)-*O*-β-D-glucop yranosyl-(1→2)-*O*-[*O*-α-L-rhamnopyranosyl-(1→4)-β-D-glucopyranosyl-(1→3)]-*O*-β-D-glucopyranosyl-(1→4)-β-D-galactopyranosyl)oxy]-5α-spirostan-12-one (furcreastatin) (**172**), (25*R*)-5α-spirostan-3β-yl-*O*-β-D-glucopyranosyl-(1→3)-*O*-β-D-glucopyranosyl-(1→2)-*O*-[*O*-α-L-rhamnopyranosyl-(1→4)-β-D-glucopyranosyl-(1→3)]-*O*-β-D-glucopyranosyl-(1→4)-β-D-galacto-pyranoside (**173**), and (25*R*)-26-[(β-D-glucopyranosyl)oxy]-22α-methoxy-5α-furostan-3β-yl-*O*-β-D-glucopyranosyl-(1→3)-*O*-β-D-glucopyranosyl-(1→2)-*O*-[*O*-α-L-rhamnopyranosyl-(1→4)-β-D-glucopyranosyl-(1→3)]-*O*-β-D-glucopyranosyl-(1→4)-β-D-galacto-pyranoside (furcreafurostatin-22-*O*-methyl ether (**174**) [98] (Figure 3.16).

3.3.2.4   Saponins from Asparagaceae

Asparagaceae is a family of flowering plants, placed in the monocots, includes 114 genera with a total of 2900 known species [99]. This family is extremely diverse, and its members are united primarily by genetic and evolutionary relationships rather than morphological similarities.

Phytochemical investigation of the aerial parts of *Chlorophytum deistelianum* led to the isolation of four previously undescribed steroidal saponins called chlorodeistelianosides A-D with five known ones. Their structures were established as (25*R*)-3β-[(β-D-glucopyranosyl-(1→3)-[α-L-rhamnopyrano syl-(1→4)]-β-D-xylopyranosyl-(1→3)-[β-D-glucopyranosyl-(1→2)]-β-D-glucopyranosyl-(1→4)-β-D-galactopyranosyl)-oxy]-5α-spirostan-12-one (chlorodeistelianoside A) (**175**), (24*S*, 25*S*)-24-[(β-D-glucopyranosyl)oxy]-3β-[(β-D-

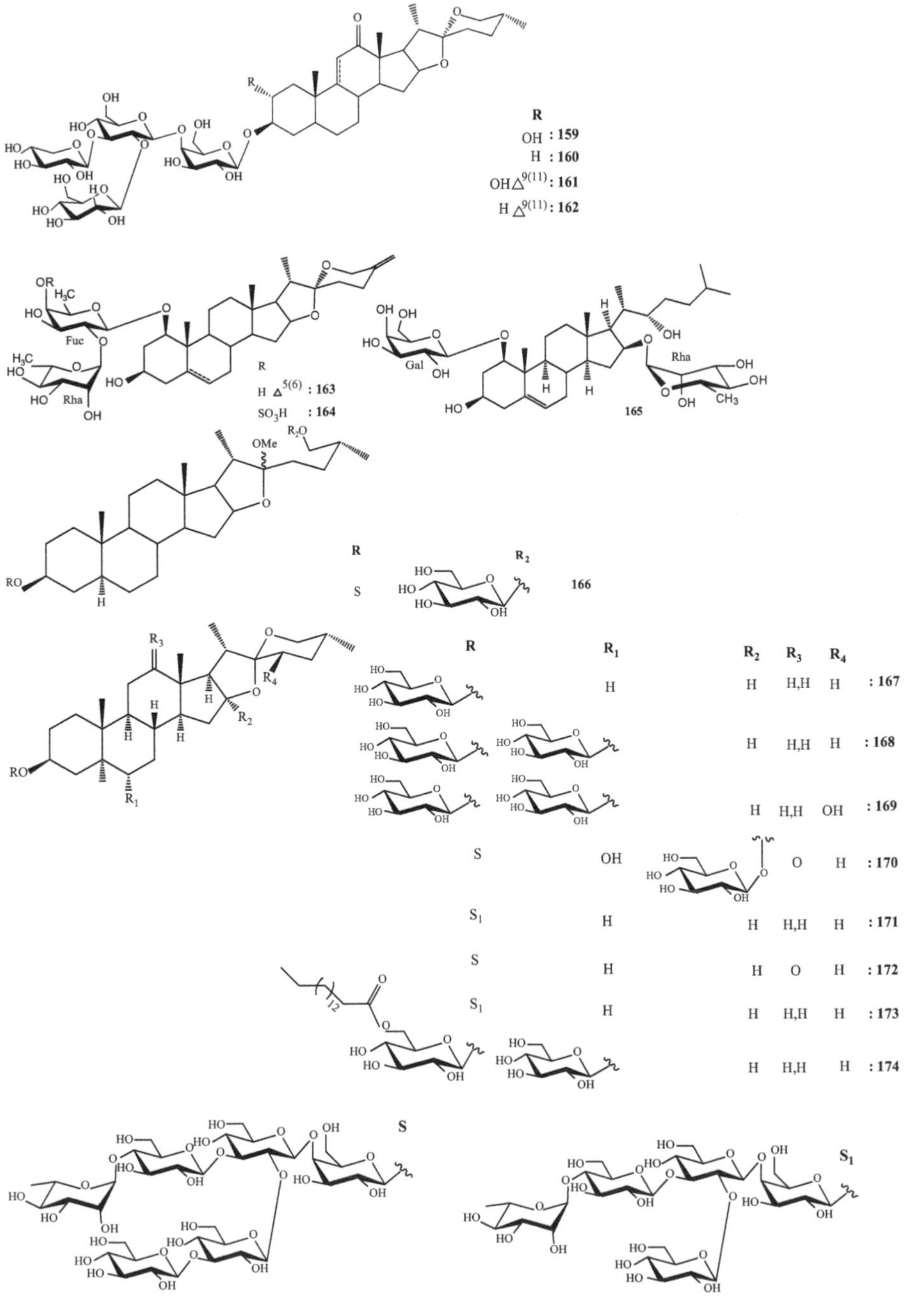

FIGURE 3.16 Saponins from *Furcraea bedinghausii*, *Cordyline fruticosa* and *Furcraea tuberosa*.

glucopyranosyl-(1→2)-[β-D-xylopyranosyl-(1→3)]-β-D-glucopyranosyl-(1→4)-β-D-galactopyranosyl)oxy]-5α-spirostan-12-one (chlorodeistelianoside B) (**176**), (25*R*)-3β-[(β-D-glucopyranosyl-(1→2)-[β-D-xylopyranosyl-(1→3)]-β-D-glucopyran osyl-(1→4)-β-D-galactopyranosyl)oxy]-5α-spirostan-12-one (**177**), (25*R*)-15α-hydro xy-5α-spirostan-3β-yl-β-D-glucopyranosyl-(1→2)-[β-D-xylopyranosyl-(1→3)]-β-D-glucopyranosyl-(1→4)-β-D-galactopyranoside (solanigroside G) (**178**), (25*R*)-2α-hydroxy-5α-spirostan-3β-yl β-D-glucopyranosyl-(1→2)-[β-D-xylopyranosyl-(1→3)]-β-D-glucopyranosyl-(1→4)-β-D-galactopyranoside (F-gitonin) (**179**)), (25*R*)-26-[(β-D-glucopyranosyl)oxy]-3β-[(β-D-glucopyranosyl-(1→2)-[β-D-xylopyranosyl-(1→3)]-β-D-glucopyranosyl-(1→4)-β-D-galactopyranosyl)oxy]-5α-furost-20(22)-en-12-one (chlorodeistelianosides D) (**180**), (25*R*)-26-[β-D-glucopyranosyl)oxy]-2α-hydroxy-22α-methoxy-5α-furostan-3β-yl   β-D-glucopyranosyl-(1→2)-[β-D-xylopy ranosyl-(1→3)]-β-D-glucopyranosyl-(1→4)-β-D-galactopyranoside (chlorodeistelia-noside C) (**181**), (25*R*)-26-[(β-D-glucopyranosyl)oxy]-22α-hydroxy-3β-[(β-D-gluco pyranosyl-(1→2)-[β-D-xylopyranosyl-(1→3)]-β-D-glucopyranosyl-(1→4)-β-D-galactopyranosyl)oxy]-5α-furostan-12-one polianthoside D (**182**), and (25*R*)-26-[(β-D-glucopyranosyl)oxy]-22α-methoxy-5α-furostan-3β-yl β-D-glucopyranosyl-(1→2)-[β-D-xylopyranosyl-(1→3)]-β-D-glucopyranosyl-(1→4)-β-D-galactopyranoside (**183**) [100] (Figure 3.17).

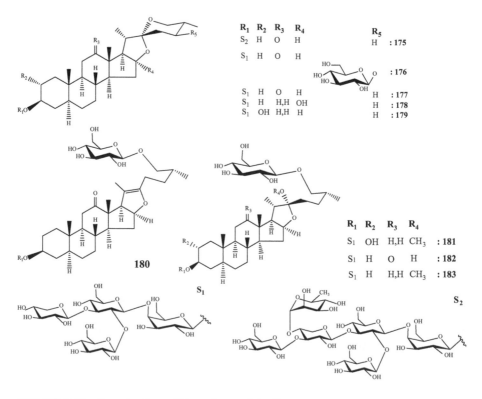

FIGURE 3.17   Saponins from *Chlorophytum deistelianum.*

3.3.2.5 Saponins from Arecaceae

Distributed mainly throughout the tropics and sub-tropics, the Arecaceae family is highly variable and exhibits a tremendous morphological diversity. Arecaceae are found in a wide range of tropical and sub-tropical ecological zones, but they are most common in the understory of tropical humid forests. Arecaceae, is a mono-phyletic group including 183 genera and 2364 species [101,102] among which the genus *Raffia*.

In tropical Africa, *Raffia* is locally used for tying and for making a wide range of products, including mats, baskets, hats, wallets, shoes, bags, fishing nets, hammocks, curtains, and textiles. The leaves are used for thatching, and the leaflets for plait-ing. *Raphia farinifera* is planted as a wayside and ornamental tree. Eight steroidal saponins (**184–191**), including one previously unreported derivative (**184**), have been isolated from the mesocarp of *Raphia farinifera* fruits by combined column chroma-tography and RP-HPLC methods [103].

3.3.2.6 Saponins from Asteraceae

The family Asteraceae is one of the largest angiosperm families distributed through-out the world, with more than 1620 genera and 23,600 species of trees, shrubs, and herbaceous plants. Asteraceae are important in herbal medicine, including *Vernonia guineensis* [104]. In Cameroon, this plant is used as an anthelmetic, antidote to poi-son, to treat malaria, jaundice, and infectious diseases [105].

One stigmastane derivative, vernoguinoside (16β, $22R;21,23S$-diepoxy-3β-O-β-D-glucopyranosyloxy-$21S$, 24-dihydroxystigmasta-8,14-dien-28-one) (**192**) has been isolated from the stem bark of *Vernonia guineensis* [105]. Thereafter, chemical investigation of the roots of the same plant afforded a new stigmastane derivative, vernoguinoside A (**193**) and three known saponins, vernoguinoside (**192**) [106], 3-O-β-D-glucopyranosyl-β-sitosterol (**194**), and 3-O-β-D-glucopyranosyl-stigmasterol (**195**) [107] (Figure 3.18).

3.4 BIOLOGICAL ACTIVITIES OF SAPONINS FROM CAMEROONIAN MEDICINAL PLANTS

Recently, interest has increased considerably in finding vegetable saponins with antioxidant properties. Research studies indicated significant anti-inflammatory and antioxidant properties of saponins which may be responsible for their antitumor properties [108]. At the present time, much attention is being paid to antioxidant substances because many pathological conditions are associated with oxidative stress and antioxidant capacity is widely used as a parameter to characterize food, medici-nal plants, and their bioactive components. Many saponins presented in this chapter have been screened for various activities and are listed below.

3.4.1 ANTIPROLIFERATIVE ACTIVITY

Pursaethosides B-D (**8–10**) were not cytotoxic when tested against HCT 116 and HT-29 human colon cancer cells [26]. Coriarioside A and B (**14 and 15**) when

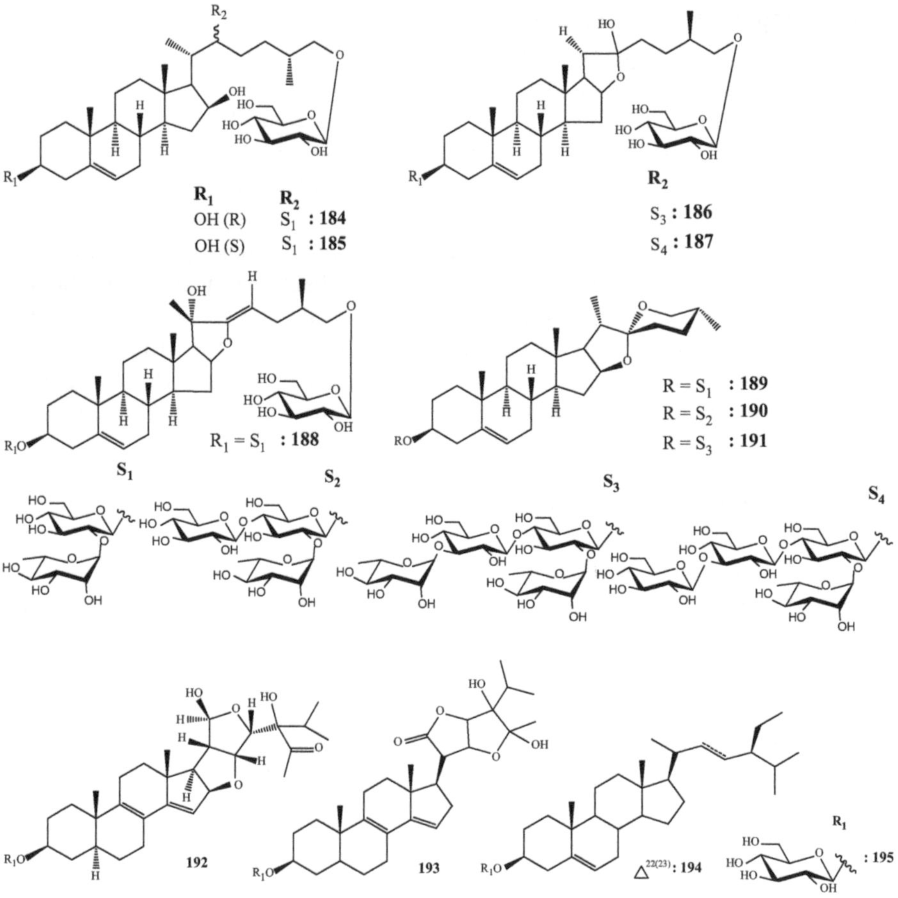

FIGURE 3.18 Saponins from *Raphia faraniferae* and *Vernonia guineensis*.

tested for cytotoxicity against two colorectal human cancer cells showed activity against the HCT 116 (IC_{50} 4.2 μM for **14** and 2.7 μM for **15**) and HT-29 (IC_{50} 6.7 μM for **14** and 7.9 μM for **15**) cell lines (29). Compounds **20** and **21** were tested for their antiproliferative activity against human glioblastoma (T98G), human squamous carcinoma (A431), human prostatic adenocarcinoma (PC3) and mouse melanoma (B16-F1) cell lines and a moderate cytotoxic potency was found for both metabolites [32]. Lebbeckosides A and B (**25** and **26**) were evaluated for their inhibitory effect on the metabolism of high-grade human brain tumor cells, the human glioblastoma U-87 MG cell lines, and the glioblastoma stem-like TG1 cells, known to be particularly resistant to standard therapies. These compounds showed significant cytotoxic activity against U-87 MG and TG1 cancer cells with IC_{50} values of 3.46 and 1.36 μM for **25**, and 2.10 and 2.24 μM for **26**, respectively [36]. Boromoenosides A-D (**27–30**) were tested for their cytotoxicity against U-87 MG human glioblastoma cell lines and TG1 glioblastoma stem-like cells with no positive activity detected [37]. The pro-apoptotic effect of the three saponins was evaluated on three human cell lines (pancreatic carcinoma AsPC-1, hematopoietic monocytic THP-1, and human fibroblast cell line BJ). Glaberrimosides A-C (**34–36**) specifically induced apoptosis of pancreatic carcinoma cell (AsPC-1) in a dose-dependent manner. More interestingly, they were inactive on monocytic (THP-1) and normal human fibroblast cell lines [44]. The apoptotic effect of zygiaosides A and B (**37** and **38**) was evaluated on the A431 human epidermoid cancer cell. Cytometric analyses showed that they induced apoptosis of human epidermoid cancer cells (A341) in a dose-dependent manner [46].

Cytotoxicity of albidosides A-I (**39–47**) was evaluated using two human cancer cells (HeLa and HL60), compounds **43** and **44** exhibited cytotoxic effect with IC_{50} 18.6 and 46.7 μM, respectively in HeLa cells and 13.3 and 15.8 μM, respectively in HL60 cells. Compound **46** exhibited strong cytotoxic effect (IC_{50} 12.7 μM on HL60 and 20.8 μM on HeLa) while compound **47** exhibited weak cytotoxicity [50]. Compounds **48–53** were investigated for their cytotoxic activity on MDA-MB 231 human breast adenocarcinoma, A375 human malignant melanoma and HCT116 human colon carcinoma cell lines. All cell lines were submitted to increasing concentrations of compounds for 72 h. IC_{50} values ranged from 40.02 to 50.12 μM for compound **53** on HCT116 and MDA-MB 231 cells, respectively, and from 66.03 to 99.05 μM for compound **52** on HCT116 and MDA-MB 231 cells, respectively. Compound **51** was inactive on the three tumor cell lines tested while **49** was slightly active against HCT116 cell line [53]. The cytotoxic activity of compounds **63** (fatsioside C1), **64** (guaianin N), **67** (chikusetsusaponin), **69,** and **70** against HCT 116 and HT-29 human tumor cell lines was determined using the MTT assay. Saponins **63** and **64** were found to be cytotoxic with $IC_{50} < 4$ mg/mL in the two cell lines, while no significant effect could be found for **67, 69,** and **70** ($IC_{50} > 100$ mg/mL) [56]. Saponins isolated from *Terminalia ivorensis* were tested for their antiproliferative activity against human breast (MDA-MB-231), prostate (PC3), colon (HCT116), and brain (T98G) cancer cell lines.

Ivorenoside A (**80**) was the most active with IC_{50} values of 3.96 and 3.43 μM against MDA-MB-231 and HCT116 cell lines, respectively compared to Cisplatin ([IC_{50}: 7.38 and 5.59 μM, respectively], and weakly active against PC3 and T98G cell lines (IC_{50} values of 16.87 and 24.14 μM, respectively). Ivorenoside A exhibited weak activity while ivorenosides B and C (**82** and **83**) were both inactive with IC_{50} values >50 μM [64]. Saponins isolated from *Butyrospermum parkii* were evaluated for their cytotoxic activities against human malignant melanoma (A375) and human glioblastoma multiforme (T98G). Parkiosides A (**84**) and B (**85**) showed *in vitro* cytotoxic activity against all the cell lines tested. Compound **86** showed better cytotoxic activity against A375 (IC_{50} 2.75 μM) and T98G (IC_{50} 2.93 μM) cell lines compared to Cisplatin (IC_{50} 7.79 and 0.50 μM, respectively) [67]. Compounds **99–104** were tested against two human colon cancer cells, HCT 116 and HT-29. In this assay, **99** and **100** showed a weak cytotoxicity (IC_{50} 24.1 and 24.0 μM, 83.0 and 83.6 μM, respectively) when tested against HT-29 and HCT 116, respectively, whereas **101–104** were not active [75].

The cytotoxicity of compounds **111**, **112** and **113** was evaluated against the HT-29 and HCT 116 human colon cancer cell lines. Compounds **111** and **112** exhibited respectively moderate and low cytotoxicity on both cell lines, with IC_{50} values in the range 7.60–70.73 μM. Compound **113** was inactive, with an $IC_{50} > 100$ μM [81].

Saponins **156–158** were evaluated for their cytotoxic activity against two human colon cancer cell lines (HCT 116 and HT-29). Some compounds isolated from *Dioscorea bulbifera* (**140–155**) were tested for cytotoxicity against urinary bladder carcinoma cells (ECV-304 cells). Results revealed a moderate activity for compounds **151**, **153** and **155** [88]. Compound **157** exhibited weak activity (IC$_{50}$ ranging from 31.0 to 48.7 µM) on both cell types and compound **158** was inactive [89].

The effect of fruticoside H-J (**163–165**) on the viability of a MDA-MB 231 human breast adenocarcinoma cell line, HCT 116 human colon carcinoma cell line, and A375 human malignant melanoma cell line was evaluated *in vitro* using the MTT assay. Fruticoside H and I (**163** and **164**) showed antiproliferative activity. IC$_{50}$ values ranged from 37.83 to 69.68 µM for 58 on A375 and MDA-MB 231cells, respectively, and from 46.59 to 59.97 µM for 59 on A375 and HCT116 cells, respectively. Fruticoside J (**165**) was inactive on the three tumor cell lines tested [95].

3.4.2 ANTIMICROBIAL AND ANTIPLASMODIAL ACTIVITIES

Compound **6** displayed antimicrobial properties against *Enterobacter aerogenes* ATCC13048, *Escherichia coli* ATCC8739, *Klebsiella pneumoniae* ATCC11296, and *Staphylococcus aureus* (MIC 8–256 µg/mL). The antibacterial and antifungal activities were in some cases equal or even higher than those of the two reference drugs thereby highlighting its potent antimicrobial activity [23]. Antimicrobial activities of **192–194** were evaluated against three bacterial species (*Salmonella typhi*, *Staphylococcus aureus* and *Shigella flexneri*) and three yeast species (*Candida albicans*, *Candida parapsilosis* and *Cryptococcus neoformans*). Compounds **192** and **193** exhibited both antibacterial and antifungal activities that varied between the microbial species (MIC 7.81–125 µg/mL), while *S. flexneri* and *C. albicans* were sensitive to all the tested compounds [107].

Compound **193** exhibited significant inhibitory activity against four strains of bloodstream trypomastigotes *Trypanosoma brucei rhodesiense* with IC$_{50}$ values in the range between 3 and 5 mg/mL [105]. It was also tested against trypanosome isolates grown in blood forms in an HMI-18 medium and IC$_{50}$ values determined after 48 h [107].

3.4.3 ANTI-INFLAMMATORY ACTIVITY

Compound **1** showed significant inhibitory activity against snake venom phosphodiesterase I [19]. Pallidiosides A-C (**3–5**) isolated from *Oxyanthus pallidus* possess analgesic effects but lack anti-inflammatory activities. Compound **3** significantly inhibited carrageenan-induced paw edema in the rats, **4** was moderately active, whereas **5** showed very weak activity [21]. The triterpene saponin β-D-glucopyranosyl 2α, 3β, 6β-trihydroxy-23-galloylolean-12-en-28-oate (**77**) obtained from the stem bark of *Combretum molle* exhibited significant activity against carrageenan-induced paw edema in rat (57.55% inhibition) while metabolites **78** and **79** showed weak activity [63].

3.4.4 ANTIOXIDANT ACTIVITY

Rheedinosides A and B (**20** and **21**) were evaluated for their antioxidant properties and moderate activity was found for both, **20** being more active than **21** [32].

Compound **80** isolated from *Terminalia ivorensis* showed better antiradical activity against the ABTS+ radicals (IC$_{50}$: 19.66 µM) compared to Trolox (IC$_{50}$: 79.5 µM) used as a reference [64]. In the nitric oxide production inhibition experiment, compounds **52** and **53** exhibited weak NO inhibitory activity (47.9 and 22.1 µM, respectively) against LPS-induced macrophage cells [52]. Parkioside B (**85**) isolated from *Butyrospermum parkii* showed good scavenging activity against 2,2-diphenyl-1-picrylhydrazyl (DPPH) (IC$_{50}$ 16.2 µM) and 2,2'-azino-bis(3-ethylbenzothiazoline-6-sulfonic acid) (ABTS) (IC$_{50}$ 25.0µM) radicals compared to Trolox (IC$_{50}$=18.8 and 12.5 µM, respectively) used as controls [67]. Leptocarposide (**97**) reveals nonpotent antioxidant activities when screened for its potential antioxidant properties by using the radical scavenging assay model 2,2-diphenyl-1-picrylhydrazyl [72].

3.5 CONCLUSION

This chapter presents 195 examples of saponins, including 106 triterpene and 89 steroidal saponins from Cameroonian medicinal plants, reported between 2003 and 2017. This survey of studies performed over the last few years has demonstrated that saponins isolated from many plants possess a broad spectrum of biological and pharmacological activities. As has been shown in many *in vitro* and *in vivo* bioassays, they exhibit antimicrobial, anti-inflammatory, antioxidant, and antiproliferative activities.

From a therapeutic point of view, saponin containing plants described here are used in traditional medicine against some infections caused by bacteria and fungi. It is hoped that this chapter may contribute to bringing a rational basis on saponins isolated from the medicinal plants. Furthermore, the information collected here will provide wider opportunities for the development and utilization of saponins.

ACKNOWLEDGMENTS

The authors gratefully acknowledge all their international collaborators who have enormously contributed to the chemical analysis and biological screening of compounds presented in this chapter, mainly professor Marie Aleth Lacaille-Dubois of the University of Burgundy, France, professor Hee Juhn Park of Sangji University, Republic of Korea, professor Melzig Matthias, Freie University of Berlin, Germany, professor Lupidi Giulio, University of Camerino, Italy and professor Tomofumi Miyamoto, Kyushu University, Japan.

REFERENCES

1. Hostettmann, K., Marston, A., Ndjoko, K. and Wolfender, J.-L., 2000. The potential of African plants as a source of drugs. *Current Organic Chemistry* 4: 973–1010.
2. Oakenfull, D. and Sidhu, S., 1986. Saponins. In *Toxicants of plant origin*; P.R. Cheebo, Ed.; CRC Press, Boca Ratton, Florida. pp. 98–143.
3. Vasilyeva, I.S. and Paseshnichenko, V.A., 2000. Steroid glycosides of the plants and the cell cultures of *Dioscorea* species: Metabolism and biological activity. *Uspekhi Biologicheskoi Khimii* 40: 153–204.

4. Bruneton, J., 1999. *Pharmacognosie; Phytochimie, Plantes médicinales*, 3eme édition, Tec et Doc, Lavoisier, Cachan Cedex, France, p. 1120.

5. Faizal, A. and Geelen, D., 2013. Saponins and their role in biological processes in plants. *Phytochemistry Reviews* 12: 877–893.

6. Dewick, M.P., 2002. *Medicinal natural products a biosynthetic approach*, 2nd edition, John Wiley & Sons, Hoboken, NJ, p. 507.

7. Vincken, J.-P., Heng, L., De Groot, A. and Gruppen, H., 2007. Saponins, classification and occurrence in the plant kingdom. *Phytochemistry* 68: 275–297.

8. Friedman, M., 2006. Potato glycoalkaloids and metabolites: roles in the plant and in the diet. *Journal of Agricultural and Food Chemistry* 54:55–81.

9. Hoyer, G.A., Sucrow, W. and Winkler, D., 1975. Diosgenin saponins from *Dioscorea floribunda*. *Phytochemistry* 14: 539–542.

10. Pohl, T.S., Crouch, N.R. and Mulholland, D.A., 2000. Southern African Hyacinthaceae: chemistry, bioactivity and ethnobotany. *Current Organic Chemistry* 4: 1287–1324.

11. Marston, A., Wolfender, J.L. and Hostettmann, K., 2000. Analysis and isolation of saponins from plant material. In *Saponins in Food, Feedstuffs and Medicinal Plants, Annual Proceedings of the Phytochemical Society*, pp. 1–12.

12. Muir, A.D., Ballantyne, K.D. and Hall, T.W., 2000. LC-MS and LC-MS/MS analysis of saponins and sapogenins - comparison of ionization techniques and their usefulness in compound identification. In *Saponins in Food, Feedstuffs and Medicinal Plants, Annual Proceedings of the Phytochemical Society*, pp. 35–42.

13. Schopke, T., 2000. Non-NMR methods for structure elucidation of saponins. In *Saponins in Food, Feedstuffs and Medicinal Plants. Annual Proceedings of the Phytochemical Society*, pp. 106–113.

14. Oleszek, W., and Bialy, Z., 2006. Chromatographic determination of plant saponins: An update (2002–2005). *Journal of Chromatography A* 1112: 78–91.

15. Agrawal, P.K., 1996. *A systematic NMR approach for the determination of the molecular structure of steroidal saponins, in: Saponins used in food and agriculture*, Springer, USA, pp. 299–315.

16. Agrawal, P.K., 1992. NMR spectroscopy in the structural elucidation of oligosaccharides and glycosides. *Phytochemistry* 31:3307–3330.

17. Scognamiglio, M., Severino, V., D'Abrosca, B., Chambery, A. and Fiorentino, A., 2015. *Structural elucidation of saponins: A combined approach based on high-resolution spectroscopic techniques*. Elsevier 45: 120. http://dx.doi.org/10.1016/B978-0-444-63473-3.00004-6.

18. Atta-ur-Rahman and Choudhary, I.C., 2002. Biodiversity-A wonderful source of exciting new pharmacophores. Further to a new theory of memory. *Pure and Applied Chemistry* 74: 511–517.

19. Fatima, N., Tapondjou, A.L., Lontsi, D., Sondengam, B.L., Atta-Ur-Rahman and Choudhary, I.C., 2002. Quinovic acid glycosides from *Mitragyna stipulosa*, first examples of natural inhibitors of snake venom phosphodiesterase I. *Natural Product Letters* 16: 389–393.

20. Piegang, B.N., Tigoufack I.B.N., Ngnokam, D., Achounna, A.S., Watcho, P., Greffrath, W. et al., 2016. Cycloartanes from *Oxyanthus pallidus* and derivatives with analgesic activities. *BMC Complementary and Alternative Medicine* 16: 97–108.

21. Nzedong, T.B.I., Ngnokam D., Tapondjou A.L., Harakat D. and Voutquenne L., 2010. Cycloartane glycosides from leaves of *Oxyanthus pallidus*. Phytochemistry 71: 2182–2186.

22. Edeoga, H.O., Okwu, D.E. and Mbaebie, B.O., 2005. Phytochemical constituents of some Nigerian plants. *African Journal of Biotechnology* 4: 685–688.

23. Chouna, J.R, Tamokou, J.D., Pépin Nkeng-Efouet-Alango, P., Lenta, N.B. and Sewald, N., 2015. Antimicrobial triterpenes from the stem bark of *Crossopteryx febrifuga*. *Zeitschrift fur Naturforschung C* 70: 169–173.

24. Hutchinzon, J. and Dalziel, J.M., 1958. *Flora of west Tropical Africa*, 2nd Edition, Vol 1, Whitefrians Press Ltd, London, pp. 484–504.
25. Palgrave, C.M., 2002. *Les arbres de l'Afrique Australe*. Troisième Edition, Struik, l'Afrique du sud.
26. Tapondjou, A.L, Miyamoto, T., Mirjolet J.F., Guilbaud, N. and Lacaille-Dubois, M.A., 2005. Pursaethosides A-E, triterpene saponins from *Entada pursaetha*. *Journal of Natural Products* 68: 1185–1190.
27. Noté, P.O., Tapondjou A.L., Mitaine-offer, A.-C., Miyamoto, T., Paululat, T.D., Pegnyemb, D.E. et al., 2009. Tetrapteroside A and B, two new oleanane-type saponins from *Tetrapleura tetraptera*. *Magnetic Resonance in Chemistry* 47: 277–282.
28. Tabuti, J.R.S. and Mugula, B.B., 2007. The ethnobotany and ecological status of *Albizia coriaria* Welw.ex Oliv. in Buwagi and Namizi parishes, Uganda. *African Journal of Ecology* 45: 126–129.
29. Noté, P.O., Mitaine-Offer, A.-C., Miyamoto, T., Paululat, T., Mirjolet, J.-F., Duchamp, O. et al., 2009. Cytoxic Acacic Acid Glycosides from the roots of *Albizia coriaria*. *Journal of Natural Products* 72: 1725–1730.
30. Noté, P.O., Chabert, P., Pegnyemb, E.D., Weniger, B., Lacaille-Dubois, M.-A. and Lobstein, A., 2010. Structure elucidation of new acacic acid-type saponins from *Albizia coriaria*. *Magnetic Resonance Chemistry* 48:829–836.
31. Kerharo J. and Adam J.G., 1974. *La pharmacopée sénégalaise traditionnelle : plantes médicinales et toxiques*. Ed. Vigot Frère, Paris, pp. 343–345.
32. Nzowa, K.L, Barboni, L., Teponno, R.B., Ricciutelli, M., Lupidi, G., Quassinti L. et al., 2010. Rheediinosides A and B, two antiproliferative and antioxidant triterpene saponins from *Entada rheedii*. *Phytochemistry* 71: 254–261.
33. Adjanohoun, J.E., Aboubakar, N., Dramane, K., Ebot, M.E., Ekpere, J.A., Enow-Orock, E. G., et al., 1996. Contribution to Ethnobotanical and Floristic Studies in Cameroon: traditional medicine andpharmacopoeia. Technical and Research Commission of Organisation of African Unity (OAU/STRC), pp. 240–241.
34. Tené, M., Chabert, P., Noté, O., Kenla, N.J.T., Tane, P. and Lobstein, A., 2011. Triterpenoid saponins from *Cylicodiscus gabunensis*. *Phytochemistry Letters* 4: 89–92.
35. Arbonnier, M., 2009. *Arbres, arbustes et lianes des zones sèche d'Afrique de l'Ouest*; MNHN Editions Quæ, Paris. pp. 383, 576.
36. Noté, P.O., Jihu, D., Antheaume, C., Zeniou, M., Pegnyemb, D.E., Guillaume, D. et al., 2015. Triterpenoid saponins from *Albizia lebbeck* (L.) Benth and their inhibitory effect on the survival of high grade human brain tumorCells. *Carbohydrate Research* 404: 26–33.
37. Noté, P.O., Jihu, D., Antheaume, C., Guillaume, D., Pegnyemb, E.D., Kilhoffer, C.M. and Lobstein, A., 2015. Triterpenoid saponins from *Albizia boromoensis* Aubrév.& Pellegr. *Phytochemistry Letters* 11: 37–42.
38. Burkill, H.M., 1995. *The useful plants of west tropical Africa. Vol. 3: Families J-L*, Royal Botanic Gardens, Kew, vol. 3. pp. 857.
39. Noumi, E. and Yomi, A., 2001. Medicinal plants used for intestinal diseases in Mbalmayo Region, Central Province, Cameroon. *Fitoterapia* 72: 246–250.
40. Betti, J.L., 2002. Medicinal plants sold in Yaoundé markets, Cameroon. *African Study Monographs* 23: 47–64.
41. Noumi, E. and Tchakonang, N.Y.C., 2001. Plants used as abortifacients in the Sangmelima region of Southern Cameroon. *Journal of Ethnopharmacology* 76: 263–270.
42. Noté, P.O., Tapondjou A.L., Mitaine-offer, A.-C., Miyamoto, T., Dong J.D., Pegnyemb, D.E. et al., 2015. Triterpenoid saponins from *Piptadeniastrum africanum* (Hook. f.) Brenan. *Phytochemistry Letters* 6: 505–510.
43. Arbonier, M., 2002. Arbres, *Arbustes et Lianes Des Zones d'Afrique de l'Ouest, Quæed*. CIRAD-MNHN, Paris.

44. Noté, P.O., Azouaou, A.S., Simo, L., Antheaume, Guillaume, D., Pegnyemb, D.E., Mullera, D.C. and Lobstein, A., 2016. Phenotype-specific apoptosis induced by three new triterpenoid saponins from *Albizia glaberrima* (Schumach.& Thonn.) Benth. *Fitoterapia* 109: 80–86.

45. Laird, S.A., Betafor, M., Enanga, M., Fominyam, C., Itoe, M., Litonga, E. et al., 1997. *Medicinal plants of the Limbe botanic garden*. Limbe Botanic Garden, Cameroon.

46. Noté, P.O., Simoa, L., Mbing, N.J., Guillaume, D., Aouazou, A.S., Muller, D.C. et al., 2016. Two new triterpenoid saponins from the roots of *Albizia zygia* (DC.) J.F. Macbr. *Phytochemistry Letters* 18: 128–135.

47. Irvine, F.R., 1961. *Woody plants of Ghana, with special reference to their uses*. Oxford University Press, London, p. 984.

48. Tijani, A.Y., Uguru, M.O. and Salawu, O.A., 2008. Antipyretic, anti-inflammatory, and anti-diarrheal properties of *Faidherbia albida* in rats. *African Journal of Biotechnology* 7: 696–700.

49. Tchoukoua, A., Tabopda, K.T., Uesugi, S., Misa, O.M., Kimura, K.-C., Eunsang K. et al., 2017. Triterpene saponins from the roots of *Acacia albida* Del. (Mimosaceae). *Phytochemistry* 136: 31–38.

50. Tchoukoua, A., Tabopda, K.T., Simo, K.I., Uesugi, S., Ohno, M., Kimura, K.-I., et al., 2017b. Albidosides H and I, two new triterpene saponins from the barks of *Acacia albida* Del. (Mimosaceae). *Natural Product Research* 32: 1–9.

51. Bouquet, A. and Debray, M., 1974. *Plantes Médicinales de la Cote d'Ivoire*. Office de la Recherche Scientifique et Technique d'Outre-mer, Paris.

52. Tapondjou, A.L., Lontsi, D., Bouda, H., Sondengam, B.L., Atta-Ur-Rahman, Choudhary, I. et al., 2002. Zemoside A, a new ursane type saponin from *Cussonia bancoensis*. *ACGC Chemical Research Communications* 15: 11–19.

53. Ponou, B.K., Nono, R.N., Teponno, R.B., Tapondjou, A.L., Lacaille-Dubois, M.-A., Quassinti, L. et al., 2014. Bafouoside C, a new triterpenoid saponin from the roots of *Cussonia bancoensis* Aubrev. & Pellegr. *Phytochemistry Letters* 10: 255–259.

54. Kougan, B.G., Miyamoto, T., Mirjolet, J.-F., Duchamp, O., Sondengam, B.L. and Lacaille-Dubois, M.-A., 2009. Arboreasides A-E, triterpene saponins from the bark of *Cussonia arborea*. *Journal of Natural Products* 72: 1081–1086.

55. Tetyana, P., Prozesky, E.A., Jäger, A.K., Meyer, J.J.M., Van Staden, J. and Van Wyk, B.-E., 2002. Some medicinal properties of *Cussonia* and Schefflera species used in traditional medicine. *South African Journal of Botany* 68: 51–54.

56. Tapondjou, A.L., Mitaine-Offer, A.C., Miyamoto, T., Lerche, H., Mirjolet, J.F., Nicolas G.N. et al., 2006. Triterpene saponins from *Schefflera abyssinica*. *Biochemical Systematics and Ecology* 34: 887–889.

57. Ndaya, T.J., Chifundera, K., Kaminsky, R., Wright, A.D. and Konig, G.M., 2002. Screening of African medicinal plants for antimicrobial and enzyme inhibitory activity. *Journal of Ethnopharmacology* 80: 25–35.

58. Njateng, G.S.S., Gatsing, Z.D.D., Donfack, A.R.N., Talla, M.F., Wabo, H.K., Tane, P. et al., 2015. Antifungal properties of a new terpernoid saponin and other compounds from the stem bark of *Polyscias fulva* Hiern (Araliaceae). *BMC Complementary and Alternative Medicine* 15: 25–45.

59. Bedir, E., Toyang, N.J., Khan, I.A., Walker, L.A. and Clark, M.A., 2001. A new dammarane-type triterpene glycoside from *Polyscias fulva*. *Journal of Natural Products* 64: 95–97.

60. Mitaine-Offer, A.C., Tapondjou A.L., Lontsi, D., Sondengam, B.L., Choudhary, M.I., Atta-Ur-Rhaman et al., 2004. Constituents isolated from *Polyscias fulva*. *Biochemical Systematics and Ecology* 32: 607–610.

61. Gill, B.S., Bir, S.S. and Singhal, V.K., 1982. Cytogenetics of some timber species of *Terminalia* Linn. (Combretaceae). *Proceedings of the Indian National Science Academy* 48: 779–790.

62. Bessong, P.O., Obi, C.L., Igumbor, E., Andreola, M.-L. and Litvak, S., 2004. *In vitro* activity of three selected South African medicinal plants against human immunodeficiency virus type 1 reverse transcriptase. *African Journal of Biotechnolology* 3: 555–559.
63. Ponou, K.B., Barboni, L., Teponno, R.B., Mbiantcha, M., Nguelefack, T.B., Park, H.-P. et al., 2008. Polyhydroxy oleanane type triterpenoids from *Combretum molle* and their anti-inflammatory activity. *Phytochemistry Letters* 1: 183–187.
64. Ponou, K.B, Teponno, R.B., Ricciutelli, M., Quassinti L., Bramucci, M., Lupidi, G. et al., 2010. Dimeric antioxidant and cytotoxic triterpenoid saponins from *Terminalia ivorensis* A. Chev. *Phytochemistry* 71: 2108–2115.
65. Kapseu, C. and Jiokap, N., 2001. Parmentier, M.; Dirand, M.D. *La Rivista Italiana DelleSostanze Grasse* 58: 31–34.
66. Ogunwande, I.A., Bello, M.O., Olawore, O.N. and Muili, K.A., 2001. Phytochemical and antimicrobial studies on *Butyrospermum paradoxum*. *Fitoterapia* 72: 54–56.
67. Tapondjou, A.L., Nyaa, T.L.B., Tane, P., Ricciutelli, M., Quassinti, L., Bramucci, M. et al., 2011. Parkioside A-C, triterpene saponins from *Butyrospermum parkii* (Sapotaceae). *Carbohydrate Research* 346: 2699–2704.
68. Wandji, J., Tillequin, F., Mulholland, D., Shirri, C.J., Tsabang, N., Seguine, E. et al., 2003. Pentacyclic triterpenoid and saponins from *Gambeya boukokoensis*. *Phytochemistry* 64: 845–849.
69. The Plant List, 2010. Version 1. Available at http://www.theplantlist.org/ (accessed 05.03.18).
70. Tapondjou, A.L., Miyamoto, T. and Lacaille-Dubois, M.A., 2006. Glucuronide triterpene saponins from *Bersama engleriana*. *Phytochemistry* 67: 2126–2132.
71. Oziegbe, M. and Faluyi, J.L., 2012. Reproductive biology of *Ludwigia leptocarpa* and L. adscendans Subsp. Diffusa in Ile Ife, Nigeria.Turk. *Journal of Botany* 36: 162–173.
72. Mabou, D.F., Foning, T.L.P., Ngnokam, D., Harakatb, D. and Voutquenne-Nazabadioko, L., 2013. Leptocarposide: a new triterpenoid glycoside from *Ludwigia leptocarpa* (Onagraceae). *Magnetic Resonance in Chemistry* 52: 32–36.
73. Mabou, D.F., Ngnokam, D., Harakatb, D. and Voutquenne-Nazabadioko, L., 2014. New oleanane-type saponins: Leputocarposide B-D, from *Ludwigia leptocarpa* (Onagraceae). *Phytochemistry Letters* 14: 159–164.
74. Masoumian, M., Arbakariya, A., Syahida, A. and Maziah, M., 2011. Flavonoids production in *Hydrocotyle bonariensis* callus tissues. *Journal of Medicinal Plants Research* 5: 1564–1574.
75. Tabopda, K.T., Mitaine-Offer, A.-C., Tanaka, C., Miyamoto, T., Tanaka, C., Mirjolet, J.-F. et al., 2012. Triterpenoid saponins from *Hydrocotyle bonariensis* Lam. *Phytochemistry* 73: 142–147.
76. Sandjo, L.P., Simo, I.K., Kuete, V., Hannewald, P., Yemloul, M., Rincheval, V. et al., 2009. Triumfettosterol Id and triumfettosaponin, a new (fatty acyl)-substituted steroid and a triterpenoid 'dimer' bis (β-D-glucopyranosyl) ester from the leaves of wild *Triumfetta cordifolia* A. Rich. (Tiliaceae). *Helvetica Chimica Acta* 92: 1748 59.
77. Ndontsa, B.L., Tchinda, A., Teponno, R.B., Mpetga, J.S., Frédérich, M. and Tane, P., 2012. Ardisikivuoside, a new triterpenoid saponin from *Ardisia kivuensis* (Myrsinaceae). *Natural Product Communications* 7: 1–2.
78. Aslam, J., Mujib, A. and Sharma, M.P., 2013. *In vitro* micropropagation of Dracaena sanderiana Sander ex Mast: An important indoor ornamental plant. *Saudi Journal of Biological Sciences* 20: 63–66.
79. Tapondjou, A.L., Ponou, K.B., Teponno, R.B., Mbiantcha, M., Djoukeng, J.D., Nguelefack, T.B. et al., 2008. *InVivo* Anti- Inflammatory effect of a new steroidal saponin, mannioside A, and its derivatives isolated from *Dracaena mannii*. *Archives of Pharmacal Research* 31: 653–658.
80. Mimaki, Y., Kuroda, M., Takaashi, Y. and Sashida, Y., 1998. Steroidal saponins from stem of *Dracaena concinna*. *Phytochemistry* 47: 1351–1356.

81. Kougan, B.G., Miyamoto, T, Tanaka, C., Paululat, T., Mirjolet, J.-F., Duchamp, O. et al., 2010. Steroidal saponins from two species of *Dracaena. Journal of Natural Products* 73: 1266–1270.

82. Antunes, A.D.S., Silva, B.P.D., Parente, J.P. and Valente, A.P., 2003. A new bioactive steroidal saponin from *Sansevieria cylindrical. Phytotherapy Research* 17: 179–182.

83. Anbu, J.S., Jayaraj, P., Varatharajan, R., John, T., Jisha, J. and Muthappan, M., 2009. Analgesic and antipyretic effects of *Sansevieria trifasciata* leaves. *African Journal of Traditional CAM* 6: 529–533.

84. Teponno, R.B., Tanaka, C., Jie, B., Tapondjou, A.L. and Miyamoto, T., 2016. Trifasciatosides A-J, steroidal saponins from *Sansevieria trifasciata. Chemical and Pharmaceutical Bulletin* 64: 1347–1355.

85. Tchegnitegni, T.B., Teponno, B.R., Siems, J.K.C., Melzig, F.M., Miyamoto, T. and Tapondjou, A.L., 2017. A dihydrochalcone derivative and further steroidal saponins from *Sansevieria trifasciata* Prain. *Zeitschrift fur Naturforschung C* 72: 477–482.

86. Caddick, L.R.P. Wilkin, P.J. Rudall, T.A.J. Hedderson, M.W. and Chase., 2002. Yams reclassified: A recircumscription of Dioscoreaceae and Dioscoreales. *Taxon* 1: 103–114.

87. Teponno, R.B., Tapondjou, A.L., Ju-Jung, H., Nam, J.-H., Tane, P. and Park, H.-J., 2007. Three new clerodane diterpenoids from the bulbils of *Dioscorea bulbifera* L. var *sativa. Helvetica Chimica Acta* 90: 1599–1605.

88. Tapondjou, A.L., Siems, J. K., Böttcher, S. and Melzig, F.M., 2013. Steroidal saponins from the flowers of *Dioscorea bulbifera* var. *sativa. Phytochemistry* 95: 341–350.

89. Tabopda, K.T., Mitaine-Offer, A.-C., Tanaka, C., Miyamoto, T., Mirjolet, J.-F., Duchamp, O., et al., 2014. Steroidal saponins from *Dioscorea preussii. Fitoterapia* 97: 198–203.

90. Teponno, R.B., Ponou, K.B., Fiorini, D., Barboni, L. and Tapondjou, A.L., 2013. Chemical constituents from the roots of *Furcraea bedinghausii* Koch. *International Letters of Chemistry, Physics and Astronomy* 11: 9–19.

91. Simmons-Boyce, J.L., Tinto, W.F., McLean, S. and Reynolds, W.F., 2004. Saponins from *Furcraea selloa* var. marginata. *Fitoterapia* 75: 634–638.

92. Elbert, L.L.J. and Roger, G.S., 1989. *Common forest trees of Hawaii (Native and introduced).* Agriculture Handbook No. 679, Washington, DC.

93. Buttner, R., 2001. *Mansfeld's encyclopedia of agricultural and horticultural crops.* Springer, Berlin.

94. Dalimartha, S., 2007. *Atlas Tumbuhan Obat Indonesia,* vol. 4. Puspa Swara, Jakarta.

95. Fouedjou, T.R., Teponno, R.B., Quassinti, L., Bramucci, M., Petrelli, D., Vitali, A.L. et al., 2014. Steroidal saponins from the leaves of *Cordyline fruticosa* (L.) A. Chev. and their cytotoxic and antimicrobial activity. *Phytochemistry Letters* 7: 62–68.

96. Smith, G.F. and Figueiredo, E., 2012. A second species of *Furcraea* Vent. (Agavaceae), *Furcrae tuberosa* (Mill.) W. T. Aiton, naturalised in South Africa. *Bradleya* 30: 107–110.

97. Alvarez de Zayas, A.M., 1996. El género *Furcraea* (Agavaceae) en Cuba. Anales del Instituto de Biologia, Universidad Nacional Autonoma, de Mexico, *Serie Botanica* 67: 329–346.

98. Tapondjou, A.L., Siems, J.K., Siems, K., Weng A. and Melzig, F.M., 2017. Flavonol glycosides and cytotoxic steroidal saponins from *Furcraea tuberosa* (Agavaceae). *Natural Product Communications* 12: 347–350.

99. Christenhusz, M.J.M. and Byng, J.W., 2016. The number of known plants species in the world and its annual increase. *Phytotaxa. Magnolia Press* 261: 201–217.

100. Tabopda, K.T., Mitaine-Offer, A.-C., Paululat, T., Stéphanie Delemasure, S., Dutartre, P., Ngadjui, T.B., et al., 2016. Steroidal saponins from *Chlorophytum deistelianum. Phytochemistry* 126: 34–40.

101. Govaerts, R. and Dransfield, J., 2005. *World checklist of palms*. Royal Botanic Gardens, Kew. UK.
102. Dransfield, J., Uhl, N.W., Asmussen, C.B., Baker, W.J., Harley, M.M. and Lewis, C.E., 2008. *Genera palmarum: The evolution and classification of palms*. Royal Botanic Gardens, Kew, UK.
103. Tapondjou, A.L., Siems J.K., Böttger, S. and Melzig F.M., 2015. Steroidal saponins from the mesocarp of the fruits of *Raphia farinifera* (Gaertn.)Hyl. (Arecaceae) and their cytotoxic activity. *Natural Product Communications* 10: 1941–1944.
104. Njateng, G.S.S., Kuiate, J.R., Gatsing, D., Tamokou, J.D., Mouokeu, R.S. and Kuete, V., 2010. Antidennatophytic activity and dermal toxicity of essential oil from the leaves of *Ageratum houstonianum* (Asteraceae). *Journal of Biological Sciences* 10: 448–454.
105. Tchinda, A.T., Tsopmo, A., Tane, P., Ayafor, J.F., Connolly, J.D. and Sterner, O., 2002. Vernoguinosterol and vernoguinoside, trypanocidal stigmastane derivatives from *Vernonia guineensis* (Asteraceae). *Phytochemistry* 59: 371–374.
106. Tchinda, A.T., Tane, P., Ayafor, J.F. and Connolly, J. D., 2003. Stigmastane derivatives and isovaleryl sucrose esters from *Vernonia guineensis* (Asteraceae). *Phytochemistry* 63: 841–846.
107. Donfack, A.R.N., Toyang, N.J., Wabo, H.K., Tane, P., Awouafack, M.D., Kikuchi, H., et al., 2012. Stigmastane derivatives from the roots of *Vernonia guineensis* and their antimicrobial activity. *Phytochemistry Letters* 5: 596–599.
108. Tundis, R., Bonesi, M., Deguin, B., Loizzo, M.R., Menichini, F., Conforti, F. and Menichini, F., 2009. Cytotoxic activity and inhibitory effect on nitric oxide production of triterpene saponins from the roots of *Physospermum verticillatum* (Waldst & Kit) (Apiaceae). *Bioorganic and Medicinal Chemistry* 17: 4542–4547.

4 Antioxidant and Immunomodulatory Activities of Cameroonian Medicinal Plants

Matthias Onyebuchi Agbo
and Patience O. Osadebe
University of Nigeria

Xavier Siwe Noundou and Rui W.M. Krause

CONTENTS

4.1 INTRODUCTION

Oxidative stress results from an imbalance between oxidized cellular components and antioxidants in favor of the oxidized components and may lead to cancer and many other metabolic diseases [1]. This stress is caused by several environmental factors such as exposure to pollutants, alcohol, medications, infections, toxins and radiation. Oxidative stress causes injury to cells, induces gene mutation and is involved in carcinogenesis by influencing intracellular signal transduction and transcription factors directly or indirectly *via* oxidants [2]. Free radicals and reactive oxygen species generated endogenously can be involved in a higher number of diseases through lipid peroxidation, protein peroxidation and DNA damage.

DOI: 10.1201/9780429506734-4

The immune system is the body's natural guard against foreign invaders and is composed of a system of biological structures, which helps to protect the body's system against many pathogens [3]. An efficient immune system detects and attacks a wide variety of agents including viruses, bacteria, fungi and parasitic worms by employing both antigen-specific (adaptive) and nonantigen-specific (innate) systems to overcome these threats [4]. Anticancer drugs and antibiotics weaken the immune system, and thus, there is growing interest in the search for natural agents that can enhance the immune system.

Immunomodulators are compounds either synthetic or natural that offer the body protection against diseases and infectious agents [5]. These compounds protect the body from immune-alteration diseases like cancer, asthma, inflammatory disorders and parasitic diseases. Synthetic immunomodulatory agents are commercially available like cyclophosphamide, but some are associated with severe-side effects like nephrotoxicity, hepatotoxicity, myelosuppression and induction of diabetics and hypertension [6]. Thus, there continues to be a search for alternative agents from natural sources with less side effects.

4.2 PROTOCOLS FOR ANTIOXIDANT *IN VITRO* ASSAY OF MEDICINAL PLANTS

There are several methods for the *in vitro* measurement of antioxidant activity of extracts or isolated compounds of natural origin. Antioxidants have several mechanisms of action, and therefore, no single assay can clearly explain the different modes of action of antioxidant behavior [7]. The assay could be (i) the cuvette assay of radical scavenging activity or (ii) the 96-well plate titer assay. The former assay method uses a UV-visible spectrophotometer to read the absorbance, while the latter method uses an ELISA plate reader to measure absorbance.

However, the commonest testing method is a chemical assay, based on the ability to scavenge various free radicals [8] using DPPH (1, 1-diphenyl-2-picrylhydrazyl) scavenging. This assay is based on the measurement of the reducing ability of antioxidants toward DPPH radicals [9]. The 2,2'-azino-bis-3-ethylbenzylthiazoline-6- sulphonic acid (ABTS) free-radical-scavenging assay is a colorimetric assay method that is based on the production of a relatively stable blue-green color when 2,2'-azinobis-(3ethyl-benzothiazoline-6-sulphonic acid) ABTS is incubated with a peroxidase (such as metmyoglobin and H_2O_2) to form a relatively stable radical cation, ABTS$^+$. The ABTS$^+$ radical formed interacts with ferric myoglobin to produce the color, measured spectrophotometrically at $\lambda_{max} = 600$ nm. Antioxidants in the fluid sample suppress this color production to a degree that is proportional to their concentrations [10]. The chelating capability for ferrous ions/ferric-reducing antioxidant power (FRAP) is a colorimetric assay that measures the ability of plasma to reduce the intense blue color of ferric tripyridyltriazine complex to its ferrous form [11] and is a widely used method to evaluate antioxidant activity. Metal chelation reduces the concentration of Fe^{2+} ion that acts as a catalyst to generate radicals, which initiates radical-mediated oxidative chain reactions in biological systems. Fe^{2+} ion is a pro-oxidant and complexes with ferrozine, thus preventing lipid peroxidation. The disadvantage of this is that the assay does not measure thiol antioxidants, such as glutathione and proteins; it measures only the reducing ability of ferric ion, which is not indicative of antioxidant activity physiologically. However, the assay method is advantageous since it is simple, speedy, inexpensive and robust and does not require specialized equipment [12]. Nitric oxide scavenging assay is based on diazotization reaction based on Griess reaction, which is an indirect assay of the nitic oxide (NO) produced [13]. The assay involves a chemical reaction utilizing sulfanilamide and naphthylethylenediamine dihydrochloride under acidic conditions. Both reagents compete for nitrite in the Griess reaction. Nitric oxide generated from sodium nitroprusside is measured by the Griess reaction. Sodium nitroprusside in aqueous solution at physiological pH spontaneously generates nitric oxide, which interacts with oxygen to produce nitric ions (NO_2^-) that can be estimated by use of Griess reagent.

4.3 IMMUNOMODULATORY ASSAY

Several *in vitro* and *in vivo* assays are available for screening the immunomodulatory activity of plant extracts or isolated natural products [14]. The humoral immunity assay/hemagglutination reaction measures the antibody levels *in vivo*. The relative strength of an antibody titer is defined as the reciprocal of the highest dilution, which is still capable of causing agglutination. The antibody titer is useful to measure the changes in the amount of the antibody during an immune response [15].

The delayed-type hypersensitivity (DTH) reaction assay measures cellular immunity *in vivo*. In the early hypersensitivity reaction, the antigen antibody forms immune complexes, which are known to induce inflammation with increasing vascular permeability, edema and infiltration of leukocytes. DTH reactions are manifestation of T cell effector arm of the immune system and are used for studying preliminary inflammatory processes [16].

In the mitogen-induced splenocyte proliferation, the lymphocyte-mediated immunity plays a key role in cellular and humoral immune responses, as the capacity to elicit an effective T- and B-lymphocyte response is ultimately dependent on the proliferative capacity of these cells [17].

4.4 *IN VITRO* PROLIFERATION ASSAY

Lymphocyte proliferation assay measures the ability of lymphocytes placed in short-term tissue culture to undergo a clonal proliferation when stimulated *in vitro* by a foreign molecule, antigen or mitogen. Lymphocyte proliferation was assessed using an MTT which involves reduction of tetrazolium salts. MTT [3-(4,5-dimethylthiazol-2-yl)-2,5-diphenyltetrazolium bromide] assay is based on the ability of living (metabolically active) cells to reduce tetrazolium salts to dark blue formazan compounds, which is largely impermeable to cell membranes unlike the dead cells. The resulting intracellular formazan can be solubilized and quantified spectrophotometrically [18]. Thus, tetrazolium salt-based colorimetric assays detect viable cells exclusively.

4.5   CAMEROONIAN PLANTS WITH ANTIOXIDANT PROPERTIES

Cassia siamea (*Fabaceae*) is a popular forestry and ornamental tree, native to South-East Asia. *C. siamea* is a medium-size tree rarely exceeding 20 m in height, more usually 10–12 m in height, but can attain 30 m under exception circumstances [19]. Flowers are bright yellow and borne in numerous large pyramidal panicles up to 60 cm long at the ends of branches. Pods are flat, 15–25 cm long, soft and ribbon-like when young, brown when ripe, indented between the seeds; there are 20–30 seeds per pod.

Centella asiatica (Gotu Kola) (*Apiaceae*) is a traditional medicine mainly renowned for its cognitive enhancing properties (usually alongside Bacopa monnieri) and its ability to regenerate wound healing and may also be anti-rheumatic. It is a herbaceous, frost-sensitive perennial plant in the flowering plant and is native to wetlands in Asia.

Cissus populnea Guill. & Perr (*Vitaceae*) is a strong woody or climbing shrub, 8–10 m long and 7.5 cm in diameter, growing in the savanna and dispersed generally throughout West Africa from the coast to the Sudanese and Sahelian woodland and South Nigeria [20]. The plant is used in the treatment of male infertility and urinary tract infections and used as a diuretic [21].

Ekebergia senegalensis A. Juss is a tree that grows up to a height of 30 feet and belongs to the *Meliaceae* family. It is commonly found in the Savannah forests. It is distributed from Senegal to Angola but can also be found in Southern and Northern parts of Nigeria. The plant and other related species are of important value in folk medicine. *E. senegalensis* and some members of its genera are claimed to be useful as an emetic in the treatment of dysentery and chronic cough and as fever remedy and for the treatment of syphilis [22]. The leaves are used by the Senegalese for curing epilepsy [23].

Eremomastax speciosa (Hochst.) Cufod (*Acanthaceae*) is an erect multi-branched tropical herb that grows in the forest. Due to its multifunctional use, it is now grown in farmlands and around domestic properties [24].

Gardenia aqualla Stapf & Hutch. is a flowering plant belonging to the family *Rubiaceae*, and it is found in the tropical and subtropical districts of Africa, including Nigeria [25]. *G. aqualla* is a savanna shrub, up to 9 feet high. Its flowers are white, turning golden-yellow later in the day and fragrant. The fruits of *G. aqualla* are oblong and woody.

Lannea kerstingii Engl. & K. Krause is a tree with a height of 12 m, usually highly branched and distributed throughout the Sudanese and Guinean Savannah [26] and belongs to the family *Anacardiaceae*. *Lannea kerstingii* Engl. & K. Krause is widely distributed in the sub-Saharan region from Senegal to Cameroon. *L. kerstingii* seeds are traditionally used in Burkina Faso as food and medicine and for skin care [27]. The roots and bark are used against diarrhea and for treatment of rachitic children and strained muscles. In the Côte d'Ivoire, the bark is used for treatment of diarrhea, edema, paralysis, epilepsy and insanity [28].

Protea elliottii (slash pine) is an evergreen conifer that has relatively long needles and grows to heights of 18–30.5 cm and with trunks averaging 61 cm [29,30]. Needles grow in cluster of 2–3 and measure approximately 31 cm in length. It has thick, plate-like bark, an extensive root system and a moderate taproot.

Terminalia macroptera is a species of flowering plant in the *Combretaceae* family and of the *Terminalia* genus. It is native to Africa, where it can be found in Benin, Burkina Faso, Ghana, Senegal, Sudan, Uganda and Nigeria. This species is used medicinally in several African countries. It is common in clear deciduous forests and in bushy and grassy savannahs. It is a small tree of 20 m in height. Its trunk can reach 40 cm in diameter, and its bark typically has deep cracks. The species can be propagated using its fruit (the *badam*, enclosing an edible almond) and with cuttings. It is used to treat infectious diseases, tuberculosis, hepatitis and dysentery [31]. Various parts of the plant are commonly used in traditional medicine against numerous diseases. Tea made from the roots offer cures for malaria, hepatitis, venereal diseases and eyesores. The roots are also used against depression, coughs, syphilis, urinary infections, female infertility, tuberculosis, snakebites and skin diseases. Juice from the bark cures earaches, and the leaves are used against tuberculosis, fever and hypertension.

Vitellaria paradoxa C. F. Gaertn. (shea nut) is a deciduous, small to medium-sized tree growing up to a height of 15–25 m belonging to the family: Sapotaceae. Leaves are caducous and spirally arranged, mostly in dense clusters at the tips of branches. The bark is conspicuously thick, corky and horizontally and longitudinally deeply fissured and offers protection to older trees against bush fires [32]. Shea nut fruits are a source of energy during the dry season. The large fleshy seeds yield about 45% edible lardlike fat, the shea nut butter, used for food and cosmetics. A by-product of the butter extraction is shea nut cake or meal, which can be used as a feedstuff [33]. The shea tree is indigenous to African savannahs from Senegal to Sudan, Western Ethiopia and Uganda. It may also be found in Cameroon, the Central African Republic, Ghana, Nigeria, Guinea and Congo. It grows on a variety of soils, such as clay, sandy clay, sand, stony soil and laterites. It prefers colluvial slopes with moderately moist deep soils rich in organic matter [33].

Dissotis perkinsiae Gilg is a shrub with height of 90–150 cm, found in tropical areas of Togo, Cameroon and Nigeria [34].

Adenocarpus mannii (Hook. f.) is a tree with yellow flowers and trifoliate leaves distributed in north tropical Africa [35,36].

Barteria fistulosa is a tree growing up to 13 m tall and is distributed in tropical regions of Nigeria and Cameroon [37]. A popular medicine within the plants' native range, the tree is harvested from the wild for local use. The stem bark, roots and leaves of *Barteria fistulosa* are widely used in baths and embrocation to treat pain and rheumatism [38]. In Gabon, a bark decoction is used to treat toothache, and in the Central African Republic, a decoction is used as nose drops and to treat headache. Young shoots are eaten as an aphrodisiac, and the powdered root is widely taken as an invigorator for men. A bark decoction is taken to treat venereal diseases and madness. In Congo, *Barteria fistulosa* is used in many formulations to treat epilepsy and snakebites. A bark decoction is used to treat smallpox and ulcerous sores.

Loranthus micranthus Linn (mistletoe) is a hemi-parasitic plant feeding on the transpiration stream of their host trees for water and mineral nutrients, even though it produces its own carbohydrates through photosynthesis [39]. Mistletoes grow on branches of trees like *Kola acuminata, Baphia nitida, Persia americana, Irvingia gabonensis, Citrus simensis, Pentaclethra macrophylla, Treculiar africana* and

Ficus exasperata where it forms pendant bushes, 2–5 feet in diameter. Its leaves are thick and leathery, oval to round, 1–2 inches long. Unlike ordinary epiphytes, mistletoes obtain all their mineral nutrients and water by forming direct connections with water-conducting veins ('*xylem*') inside the branches of host trees.

The mistletoe leaves are used in the traditional folkloric medicine of Nigeria for the management of diarrhea, epilepsy, hypertension and rheumatism [8].

The Cameroonian plants with antioxidant properties and their medicinal uses are summarized in Table 4.1, respectively, together with ethnomedicinal uses. The pictorial display of these plants is displayed in Figure 4.1.

TABLE 4.1
Cameroonian Medicinal Plants with Antioxidant Activities

S/N	Plant	Part	Family	Uses	Reference
1.	*Senna siamea*	Leaf, fruit	*Fabaceae*	Antimalaria, treatment of intestinal worms and convulsion	[40,41]
2.	*Centella asiatica*	Whole plant	*Apiaceae*	Wound healing and treatment of skin diseases	[42]
3.	*Cissus populnea*	Root	*Vitaceae*	Treatment of male infertility and urinary tract infections	[20]
4.	*Ekebergia senegalensis*	Bark	*Meliaceae*	Treatment of ovarian cyst	[43]
5.	*Eremomastax speciosa*	Leaf	*Acanthaceae*	Treatment of pains, dermatitis, anaemia, haemorrhoids and urinary tract infection	[44]
6.	*Gardenia aqualla*	Leaf	*Rubiaceae*	Treatment of impotence	[43]
7.	*Lannea kerstingii*	Bark	*Anacardiaceae*	Antidiarrhea	[26]
8.	*Protea elliottii*	Bark		Treatment of carious teeth, haemorrhoids	[43]
9.	*Terminalia macroptera*	Root	*Combretaceae*	Treatment of hepatitis	[45]
10.	*Vitellaria paradoxa*	Bark	*Sapotaceae*	Hypertension and fever	[43]
11.	*Dissosotis perkinsiae*	Leaf	*Melastomataceae*	Treatment of skin diseases and malaria	[46]
12.	*Adenocarpius mannii*	Aerial Part	*Fabaceae*	Treatment of malaria and infectious diseases	[47]
13.	*Bartera fistulosa*	Bark	*Passifloraceae*	Treatment of wounds, fever and rheumatism	[48]
14.	*Loranthus micranthus*	Leaf	*Loranthaceae*	Treatment of cancer, rheumatism and epilepsy and hypertension	[49]

FIGURE 4.1 Cameroonian medicinal plants with antioxidant properties.

4.6 SURVEY OF CAMEROONIAN PLANTS WITH IMMUNOMODULATORY PROPERTIES

Caucalis melanantha (Hocht.) is a soft-leaf herb up to 2 feet high commonly found under tall eucalyptus trees where there is no grazing. Family: *Apiaceae*

Ocimum gratissimum is an herb that grows mainly in India, Africa and parts of South-East Asia. Used in making anti-bacterial medicines. Family: *Lamiaceae*. The plant produces essential oil from its leaves and stems, which is extracted to make several medicinal substitutes.

Asystasia intrusa is a small, branching herb with ovate or ovate-lance-shaped leaves, found along jungle edges and wasteland. The leaves are whitish with a hint

FIGURE 4.2 Medicinal plants of Cameroonian origin with immunomodulatory activity. **A** = *Adenocarpus mannii*, **B** = *Caucalis melanantha*, **C** = *Ocimum gratissimum*, **D** = *Asystasia intrusa*, **E** = *Clematis chinensis* and **F** = *Loranthus micranthus parasitic on Kola acuminata*.

of purple. Commonly *Asystasia* has now become a garden plant, although in some countries it is considered a weed.

These Cameroonian immunomodulatory plants are shown in Figure 4.2.

4.7   REPORTED ANTIOXIDANT ACTIVITIES OF CAMEROONIAN MEDICINAL PLANTS

The antioxidant properties of 11 Cameroonian medicinal plants using DPPH and nitric oxide radical scavenging ability have been reported [50]. The result of the antioxidant assays showed that the medicinal plants possess radical scavenging ability except *C. spectabilis* with their IC_{50} values ranging from 14.2 ± 0.27 to $454.0 \pm 2.54 \mu g/mL$ (DPPH assay) and 205.00 ± 0.90 to $409.0 \pm 2.00 \mu g/mL$ (nitric oxide scavenging ability) as shown in Table 4.2. The DPPH assay of the medicinal plants *Cissus populnea*, *Ekebergia senegalensis*, *Protea elliottii*, *Terminalia macroptera* and *Vitellaria paradoxa* showed more radical scavenging activity than the gallic acid based on IC_{50} values. Thus, these medicinal plants could serve as sources of 'lead' compounds for anti-oxidative agents for the treatment of disease caused by oxidative stress like cancer, inflammatory disorders and diabetics. The nitric oxide assay exhibited poor radical scavenging activity than the DPPH assay. Polyphenolic compounds like flavonoids and tannins account for the antioxidant properties of plant extracts. Polyphenolic compounds act as antioxidants due to their ability to scavenge free radicals and chelate metals *in vitro* and *in vivo* [51]. Studies showed that the extract of *Vitellaria paradoxa*

TABLE 4.2

Antioxidant Activity of the Methanol Extracts of Selected Cameroonian Medicinal Plants [50]

	Antioxidant Activity	
	DPPH	Nitic Oxide
Plant Extract	$(IC_{50\,\mu g/mL})$	$(IC_{50\,\mu g/mL})$
Senna siamea	236.70 ± 3.80	NA
Centella asiatica	NA	NA
Cissus populnea	15.72 ± 1.20	409.0 ± 2.0
C. spectabilis	NA	NA
Ekebergia senegalensis	15.83 ± 0.40	299.0 ± 3.0
Eremomastax speciose	454.0 ± 2.54	NA
Gardenia aqualla	105.9 ± 2.10	278.0 ± 1.56
Lannea kerstingii	34.4 ± 2.20	253.0 ± 2.0
Protea elliottii	14.2 ± 0.27	306.0 ± 2.50
Terminalia macroptera	19.9 ± 0.70	205.0 ± 0.90
Vitellaria paradoxa	22.14 ± 0.39	290.0 ± 2.40
Standard drug	22.67 ± 0.40	108.0 ± 2.0

NA = Not active; Standard drug = Gallic acid (DPPH) and Ascorbic acid (Nitic oxide).

contains flavonoids [52], *Cissus populnea* contains tannins [53] and *Ekebergia sen-egalensis* contains tannins and flavonoids [22]. This could explain the higher radical scavenging abilities of these plant extracts compared to gallic acid.

Ndjateu et al. [47] reported the antioxidant activities of an ethanol extract of *Dissotis perkinsiae, Adenocarpus mannii* and *Barteria fistulosa* using the DPPH assay method. The radical scavenging ability of *B. fistulosa* was highest ($IC_{50} = 100.16\,\mu g/mL$) followed by *D. perkinsiae* ($IC_{50} = 130.66\,\mu g/mL$) with *A. mannii* having the least radical scavenging ability ($IC_{50} = 361.30\,\mu g/mL$) compared to the positive control (ascorbic acid; $IC_{50} = 2.41\,\mu g/mL$). Isolation of polyphenols from *A. mannii* has been reported; a cyanogenic glycoside has been isolated from *B. fistulosa* [54,55], while flavonoids have been isolated from *Adenocarpus* genus [56]. The presence of these phenolic metabolites from these plants could be responsible for the antioxidant activity of the extracts. The antioxidant activities of polyphenols isolated from *Loranthus micranthus* Linn parasitic on *Kola acuminata* and *Hevea brasiliensis* using the DPPH assay model has been reported ([57,58]. The compounds showed higher radical scavenging ability compared to the reference drug (ascorbic acid and propyl gallate).

4.8 ANTIOXIDANT ACTIVITIES OF SOME ISOLATED COMPOUNDS FROM CAMEROONIAN MEDICINAL PLANTS

Semi-preparative HPLC separation of the ethyl acetate and *n*-butanol soluble fractions of a methanol extract of the leaves of *Loranthus micranthus* Linn. parasitic on *Hevea brasiliensis* led to the isolation of 14 polyphenols with antioxidant activity

[2,8]. The antioxidant potentials of the isolated compounds (**1–14**) (IC_{50} = 23.8–50.10 µM) were evaluated using the DPPH model with propyl gallate as a positive control (IC_{50} = 67.9 µM) (Figure 4.3).

Chromatographic purification of the ethyl acetate fraction of the leaves of *Loranthus micranthus* Linn parasitic on *Kola acuminate* led to the isolation of three

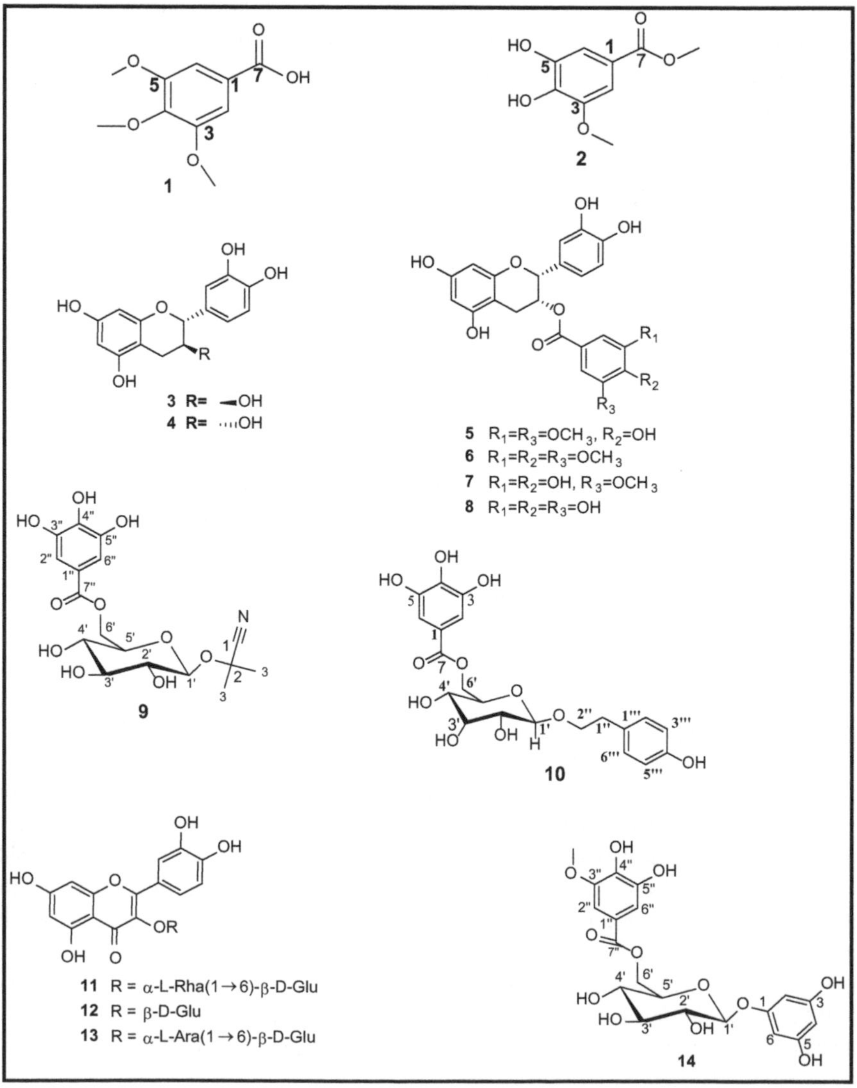

FIGURE 4.3 Compounds isolated from *Loranthus micranthus* Linn. parasitic on *Hevea brasiliensis* with antioxidant activity. 3, 4,5-tri-O-methylgallic acid (**1**), methyl 3-O-methylgallate (**2**), catechin (**3**), epicatechin (**4**), epicatechin 3-O-(3,5-O-dimethyl) gallate (**5**), epicatechin 3-O-(3,4,5-O-trimethyl) gallate (**6**), epicatechin 3-O-(3-O-methyl) gallate (**7**), epicatechin 3-O-gallate (**8**), linamarin gallate (**9**), 6-O-galloyl salidroside (**10**), rutin (**11**) and quercetin 3-O-β-D-glucopyranoside (**12**) while peltatoside (**13**) and walsuraside B (**14**) (2, 8).

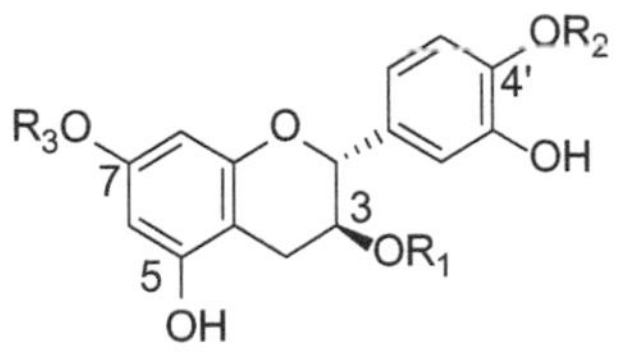

15. $R_2 = R_3 = H$; $R_1 = \alpha$-L-Rham

16. $R_1 = R_2 = H$; $R_3 = \alpha$-L-Rham

17. $R_1 = H$; $R_2 = CH_3$; $R_3 = \alpha$-L-Rham

FIGURE 4.4 Compounds isolated from *Loranthus micranthus* Linn parasitic on *Kola acuminate* with antioxidant and immunomodulatory activities. Catechin-7-*O*-rhamnoside (**15**), catechin-3-*O*-rhamnoside (**16**) and 4′-methoxy-7-*O*-rhamnoside (**17**) (57).

catechin-*O*-rhamnosides: catechin-7-*O*-rhamnoside (**15**), catechin-3-*O*- rhamnoside (**16**) and 4′-methoxy-7-*O*-rhamnoside (**17**). The antioxidant activities of isolated compounds were evaluated using DPPH assay with ascorbic acid as a positive control [57] (Figure 4.4).

The compounds (**15–17**) showed good antioxidative activity with $EC_{50} \leq 55.4 \pm 0.10$ mg/mL compared to that of the positive control (17.6 ± 1.8 mg/mL). Column chromatographic purification of the ethanol extracts of *Dissotisper kinsiae*, *Adenocaropus mannii* and *Barteria fistulosa* with gradient elution of EtOAc in *n*-hexane and EtOAc in methanol, followed by gel chromatography of the fractions obtained, yielded ten compounds (**18–27**). The antioxidant assay of some of the isolated compounds (**20, 22, 23–27**) was evaluated using DPPH model [47]. Quercetin 3-O-(6″-O-galloyl)-β-galactopyranoside (**20**) and ellagic acid (**22**) had high radical scavenging activity with IC_{50} of 9.99 ± 0.02 and 9.84 ± 0.06 µg/mL, respectively, followed by isovitexin (**25**) with an IC_{50} value of 24.27 ± 0.01 µg/mL. The introduction of a galloyl moiety in **20** and the presence of *ortho*-hydroxyls and α, β-unsaturated ketone groups in **20** and **22** improved their antioxidant potential (58). Conversely, chrysin 7-*O*-β-D-glucopyranoside (**24**), hederagenin (**26**) and shanzhiside methyl ester (**27**) had the weakest radical scavenging activity, with IC_{50} values of 484.49 ± 0.04, 234.96 ± 0.02 and 106.29 ± 0.04 µg/mL, respectively (Figure 4.5).

4.9 IMMUNOMODULATORY ACTIVITIES OF CAMEROONIAN MEDICINAL PLANTS

The immunomodulatory potentials of three catechin-*O*-rhamnosides: catechin-7-*O*-rhamnoside (**15**), catechin-3-*O*-rhamnoside (**16**) and 4′-methoxy-7-*O*-rhamnoside (**17**) isolated from *Loranthus micranthus* Linn parasitic on *Kola acuminata* on C57BL6 mice splenocytes were determined using flow cytometry techniques and compared to lipopolysaccharide (LPS; 10 µg/mL) and concanavalin A (ConA; 2 µg/mL) (57). The compounds at 100 µg/mL showed statistically significant ($p < 0.05$) stimulatory activity of $91.5\% \pm 0.2\%$, $95.2\% \pm 0.01\%$ and $94.2\% \pm 0.1\%$, respectively,

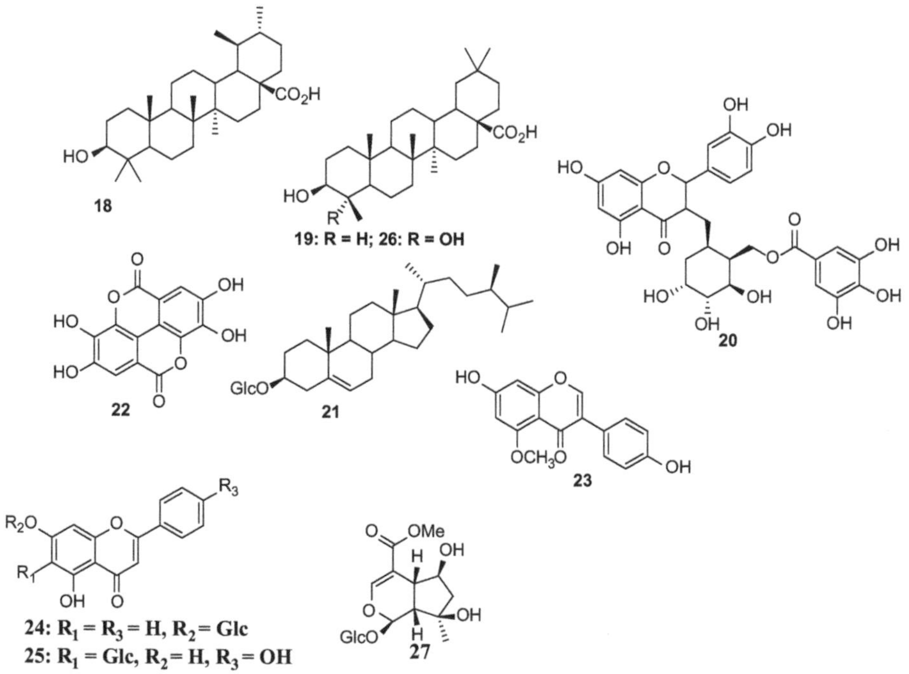

FIGURE 4.5 Chemical structures of compounds isolated from the leaves of *Dissotisper kinsiae* (18–22), *Adenocaropus mannii* (19,23–25) and the stem barks of *Barteria fistulosa* (19,21,26–27) with antioxidant activity. Ursolic acid (**18**), oleanolic acid (**19**), quercetin 3-O-(6″-O-galloyl)-β-galactopyranoside (**20**), 3-O-β-D-glucopyranoside of sitosterol (**21**), ellagic acid (**22**), isoprunetin (**23**), chrysin 7-O-β-D-glucopyranoside (**24**), isovitexin (**25**), hederagenin (**26**), shanzhiside methyl ester (**27**) (47).

compared to 2.65% ± 0.33% for the unstimulated control. However, the CD69 expression assay showed only moderate stimulation. The immunomodulatory potentials of the lupane triterpenoid esters: 7β, 15α-dihydroxyl-lup-20(29)-ene-3β-palmitate (**28**), 7β, 15α-dihydroxyl-lup-20(29)-ene-3β-stearate (**29**) and 7β, 15α-dihydroxyl-lup-20(29)-ene-3β-decadecanoate (**30**) isolated from the *n*-hexane fraction of the Eastern Nigeria mistletoe parasitic on *Kola accuminata* by the cell-mediated DTH reaction model in mice has been reported [59,60]. Only compound **28** was subjected to cell proliferation studies using C57Bl/6 splenocytes at dose levels of (5, 25 and 100 μg/mL) in the presence of controls. Also, the effect of this compound on IL-8 receptor expression was evaluated at three doses (1, 5 and 10 μg/mL) using the real-time polymerase chain reaction assay. This triterpenoid ester produced some enhancement of the splenocytes at the tested doses and at doses higher than 5 μg/mL caused inhibition of the IL-8 receptor expression, thus supporting the ethnomedicinal use of the Eastern Nigeria mistletoe in the management of diseases affecting the immune system. In addition, the immunostimulatory activity of lupinine alkaloid (**31**) (10, 25 and 100 μg/mL) isolated from *Loranthus micranthus* parasitic on *Kola accuminata* on C57BL/6 mice splenocytes using concanavalin A (2 μg/mL) as standards have been

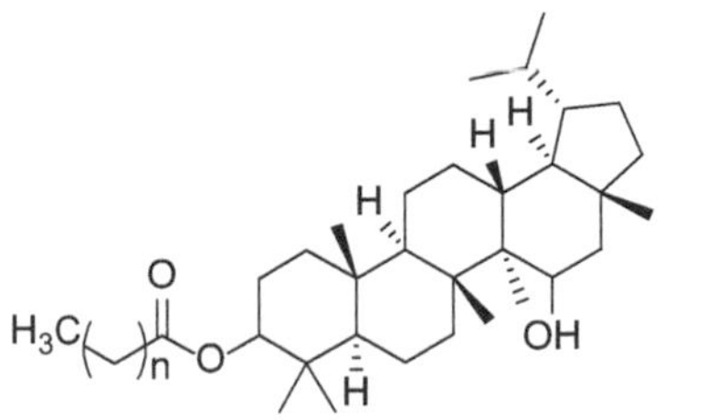

28: n = 12; 29: n = 14; 30: n = 16

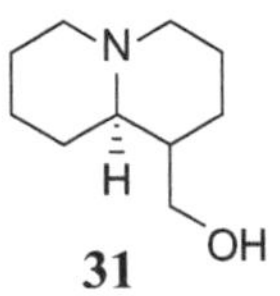

FIGURE 4.6 Chemical structures of compounds isolated from the leaves of *Loranthus micranthus* Linn parasitic on *Kola acuminate* with immunomodulatory activity. 7β 15α-dihydroxyl-lup-20(29)-ene-3β-palmitate (**28**), 7β, 15α-dihydroxyl-lup-20(29)-ene-3β-stearate (**29**) and 7β, 15α-dihydroxyl-lup-20(29)-ene-3β-decadecanoate (**30**), lupinine (**31**) (59–61).

reported. Flow cytometry techniques was used for the assay and the result compared to lipopolysaccharide (10 μg/mL) [61].

Compounds **32** at 25 μg/mL showed statistically significantly ($p < 0.05$) stimulatory activity on the splenocytes with values of $56.3\% \pm 0.3\%$ and $69.8\% \pm 0.2\%$, respectively, compared to $7.6\% \pm 0.4\%$ recorded for the control. Similarly, the CD69 expression assay showed immunostimulation with statistically significant values ($p < 0.05$) of $2.3\% \pm 0.1\%$ and $2.7\% \pm 0.03\%$, respectively, compared to $1.7\% \pm 0.1\%$ recorded for the control. This suggests that the isolated compounds possess immuno-modifying abilities (Figure 4.6). The immunomodulator potentials of the methanol extracts of mistletoe sourced from five host trees have been reported [62].

REFERENCES

1. Boudjeko, T., Megnekou, R, Woguia, A.L., Kegne, F.M., Ngomoyogoli, J.E.K., Tchapoum, C.D.N. and Koum, O., 2015. Antioxidant and immunomodulatory properties of polysaccharides from *Allanblackia floribunda* Oliv stem bark and *Chromolaena odorata* (L.) King and H.E. Robins leaves. *BMC Res Notes* 8: 759.
2. Agbo, M.O., Ezealisiji, K.M., Parker, J.E. and Obonga, W.O., 2015. Gallic Acid Derivatives (GADs) from *Loranthus micranthus* Linn parasitic on *Hevea brasiliensis* with antioxidative capacity. *Dhaka Univ J Pharm Sci* 14(2): 135–140.
3. Jong, S.C., Birmingham, J.M. and Pai, S.H., 1983. Immunomodulatory substances of fungal origin. *J Immunol Immunopharmacol* 11: 115–122.
4. Ganju, L., Karan, D., Chanda, S., Srivastava, K. and Sawhney, R., 2003. Immuno-modulatory Effects of agents of plant origin. *Biomed Phamacother* 57: 296–300.
5. Afolayan, F.I.D., Erinwusi, B. and Oyeyemi, O.T., 2018. Immunomodulatory activity of curcumin-entrapped poly D, L-lactic-co-glycolic acid nanoparticles in mice. *Integr Med Res* 7(2): 168–175. Available online at www.sciencedirect.com.Journal.
6. Waldmann, H., 2003. The new immunosuppression. *Curr Opin Chem Bio* 7: 476–480.
7. Badarinath, A.V., Rao, K.M., Chetty, C.M.S., Ramkanth, S., Rajan, T.V.S. and Gnanaprakash, K., 2010. A review on *In-vitro* antioxidant methods: Comparisons, correlations and considerations. *Int J PharmTech Res* 2(2): 1276–1285.
8. Agbo, M.O., Lai, D., Okoye, F.B.C., Osadebe, P.O. and Proksch, P., 2013. Antioxidative polyphenols from Nigerian mistletoe *Loranthus micranthus* (Linn.) parasitizing on *Hevea brasiliensis*. *Fitoterapia* 86: 78–83.

9. Prior, R.L., Wu, X. and Schaich, K., 2005. Standardized methods for the determination of antioxidant capacity and phenolics in foods and dietary supplements. *J Agric Food Chem* 3: 4290–4302.

10. Cano, A., Acosta, M. and Arnao, M.B., 2000. A method to measure antioxidant activity in organic media: application to lipophilic vitamins. *Redox Rep* 5: 365–370.

11. Benzie, I.F. and Strain, J.J., 1999. Ferric reducing antioxidant power assay; direct measure of total antioxidant activity of biological fluids and modified version for simultaneous measurement of total antioxidant power and ascorbic acid concentration methods. *Enzymol* 299: 15–27.

12. Huang, D., Ou, B. and Prior, R.L., 2005. The chemistry behind antioxidants capacity assays. *J Agric Food Chem* 53(6): 1841–1856.

13. Bryan, N.S, Grisham, M.B, 2007. Methods to direct nitric oxide and its metabolites in biological samples. *Free Radic Bio Med.* 43(5): 645–657.

14. Akerkar, A.S., Ponkshe, C.A. and Indap, M.M., 2009. Evaluation of immunomodulatory activity of extracts from marine animals. *Indian J Mar Sci* 38(1): 22–27.

15. Pradhan, D., Panda, P.K. and Tripathy, G., 2009. Evaluation of the immunomodulatory activity of the methanolic extract of *Couroupita guianesis* Aubl. Flower in rats. *Nat Prod Radiance* 8(1): 37–42.

16. Rajeswari, G., Priyanka, B., Amrutha, R.E., Rajaram, C., Kanhere, R.S. and Kumar, S.N., 2013. *Hibiscus titiaceus*: A possible immunomodulatory agent. *J Pharm Res* 6: 742–747.

17. Sghaier, M.B., Krifa, M., Bhouri, R.M.W., Ghedira, K. and Chekir-Ghedira, L., 2011. *In vitro* and *in vivo* immunomodulatory and anti-ulcerogenic activities of *Teucrium ramosissimum* extracts. *J Immunotoxicol* 8(4): 288–297.

18. Mosmann, T., 1983. Rapid colorimetric assay for cellular growth and survival: application to proliferation and cytotoxicity assays. *J Immunol Methods* 65: 55–63.

19. Heinsleigh, T.E. and Holaway, B.K., 1988. *Agroforestry species for the Philippines*. Metro Manila Philippines: US Peace Corps, AJA Printers.

20. Soladoye, M.O. and Chukwuma, E.C., 2012. Quantitative phytochemical profile of the leaves of *Cissus populnea* Guill. & Perr. (*Vitaceae*)-an important medicinal plant in Central Nigeria. *Arch Applied Sci Res* 4(1): 200–206.

21. Ojekale, A.B., Lawal, O.A., Jewo, P.I., Oguntola, J.A. and Abdu, L.O., 2015. *Cissus populnea* (Guill & Perr): A study of the aqueous extract as potential spermatogenic enhancers in male wistar rats. *Am J Med Biol Res* 3(5): 124–127.

22. Ndukwe, I.G., Habila, J.D., Bello, I.A. and Adeleye, E.O., 2006. Phytochemical analysis and antimicrobial screening of crude extracts from the leaves, stem bark and root bark of *Ekebergia senegalensis* A. Juss. *Afr J Biotechnol* 5(19): 1792–1794.

23. Dalziel, J. M. and Hutchinson J., 1937. *The Useful Plants of west Tropical Africa*. Crown Agent for the Colonies. London, pp. 318–330.

24. Ndem, J.I., Otitoju, O., Akpanaiabiatu, M.I., Uboh, F.E., Uwah, A.F. and Edet, O.A., 2013. Haematoprotective property of *Eremomastax speciosa* (Hochst.) on experimentally induced anaemic Winstar rats. *Ann Biol Res* 4(6): 356–360.

25. Ivan, O.C., Hitler, L., Joseph, J., Oyetola, O., Udochukwu, A.O., Maraga, T.N. and Isa, P.A., 2018. Phytochemical screening and corrosion inhibition of the ethanolic leave extracts of *Gardenia aqualla* Stapf & Hutch in 1M H_2SO_4 acid solution. *Am J Appl Chem* 6(1): 1–5.

26. Njinga, N.S., Sule, M.I., Pateh, U.U., Hassan, H.S., Usman, M.A. and Haruna, M.S., 2013. Phytochemical and antidiarrhea activity of the methanolic extract of the stem bark of *Lannea kerstingii* (*Anacardiaceae*). *J Nat Prod Plant Resour* 3(3): 43–47.

27. Ouilly, J.T., Bazongo, P., Bougma, A., Kaboré, N., Lykke, A.M., Ouédraogo, A. and Bassolé, I.H.N., 2017. Chemical composition, physicochemical characteristics, and nutritional value of *Lannea kerstingii* seeds and seed oil. *J Anal Methods Chem*; Article ID 2840718, 6 pages.

28. AbdulRahaman, A.A., Kolawole, O.S. and Oladele, F.A., 2014. Leaf epidermal features as taxonomic characters in some *Lannea spieces* (*Anacardiaceae*) from Nigeria. *Phytologia Balcanica* 20(2–3): 227–231.

29. Harlow, W.M., Harrar, E.S., Ellwood, S. and White, F.M., 1979. *Textbook of dendrology*. 6th edition. McGraw-Hill, Inc. New York. p. 510.

30. Anderson, R., 1998. *Guide to Florida's Trees (Guide to Florida Wildlife and Nature)*. Winner Enterprises, USA, p. 68.

31. Ibrahim, H., 2005. Pharmacognostic studies of the leaves of *Terminalia macroptera*. *Niger J Nat Prod Med* 9: 14–18.

32. Orwa, C., Mutua, A., Kindt, R., Jamnadass, R. and Simons, A., 2009. Agroforestree Database: A tree reference and selection guide version 4.0. Available online at https://www.worldagroforestry.org/publication/agroforestree-database-tree-species-reference-and-selection-guide-version-40

33. Nikiema, A, and Umali B.E, 2007. *Vitellaria paradoxa* C.F. Gaertn. Record from Protabase. In van der Vossen, H.A.M. and Mkamilo G.S. (eds.). PROTA (Plant Resources of Tropical Africa/Ressources végétales de l'Afrique tropicale), Wageningen, Netherlands.

34. Hutchinson, J. and Dalziel, J., 1954. *Melastomataceae*. Flora of West Tropical Africa. Crown Agents for Oversea Governments and Administrations, London. p. 295.

35. Cubas, P., Pardo, C., Tahiri, H. and Castroviejo, S., 2010. Phelogeny and evolutionary diversification of *Adenocarpus* DC. (*Leguminosae*). *Taxon* 59: 720–732.

36. Burrows, J.E. and Willis, C.K., (eds.) 2005. Plants of the Nyika Plateau. Southern African Botanical Diversity Network Report No. 31, SABONET, Pretoria. p.151.

37. Vanderplank, J., 1996. *Passion Flowers*. Cassell Pub. Ltd., London.

38. Neuwinger, H.D., 2000. *African traditional medicine: A dictionary of plant uses and applications*. Medpharm Scientific, Stuttgart, Germany. p. 589.

39. LuÈttge, U., Mundayatan, H.G., de Mattos, E.A., Trimborn, P., Franco, A.C., Caldas, L.S. and Ziegler, H., 1998. Photosynthesis of mistletoes in relation to their hosts at various sites in tropical Brazil. *Trees* 12: 167–174.

40. Lose, G.A., Bernard, S.J. and Leihner, D.E., 2000. Studies on agro forestry hedgerow system with *Senna siamea* rooting patterns and competition effects. *J Sci* 38: 57–60.

41. Alli-Smith, Y.R., 2009. Determination of chemical composition of *Senna siamea* (cassia leaves). *Pak J Nutr* 8(2): 119–121.

42. Zainol, N.A., Voo, S.C., Sarmidi, M.R. and Aziz, R.A., 2008. Profiling of *Centella asiatica* (L.) urban extract. *Malaysian J Anal Sci* 12(2): 322–327.

43. Jiofack, T., Fokunang, C., Guedje, N., Kemeuze, V., Fongnzossie, E., Nkongmeneck, B.A., et al., 2010. Ethnobotanical uses of medicinal plants of two ethnoecological regions of Cameroon. *Int J Med Med Sci* 2(3): 60–79.

44. Oben, J.E., Sheila, E.A., Gabriel, A.A. and Musoro, D.F., 2006. Effect of *Eremomastax speciosa* on experimental diarrhea. *Afr J Trad* 3: 95–100.

45. Diniz, M.A., Silva, O., Paulo, M.A. and Gomes, E.T., 1996. Medicinal uses of plants from Guinea-Bissau. In: Van der Maesen LJG, van der Burgt XM, Van Medenbach de Rooy JM, editors. *The biodiversity of African plants*. Kluwer Academic Publishers, Dordrecht, p. 727–731.

46. Haxaire, C., 1979. Phytothérapie et Médecine Familial chez Gbaya-Kara (République Centrafricaine). Thése de doctorat Université de Paris, Fac. Pharmacie. p.320.

47. Ndjateu, F.S.T., Tsafack, R.B.N., Nganou, B.K., Awouafack, M.D., Wabo, H.K., Tene, M., Tane, P. and Eloff, J.N., 2014. Antimicrobial and antioxidant activities of extracts and ten compounds from three Cameroonian medicinal plants: *Dissotis perkinsiae* (*Melastomaceae*), *Anenocarpus mannii* (Fabaceae) and *Barteria fitulosa* (*Passifloraceae*). *S Afr J Bot* 91: 37–42.

48. Gangoué-Piéboji, J., Eze, N., Djintchui, A.N., Ngameni, B., Tsabang, N., Pegnyemb, D.E., Biyiti, L., Ngassam, P., Koulla-Shiro, S. and Galleni, M., 2009. The *in vivo*

antimicrobial activity of some traditionally used medicinal plants against beta-lactam-resistant bacteria. *J Infect Dev Countries* 3: 671–680.

49. Griggs, P., 1991. Mistletoe, myth, magic and medicine. *The Biochem* 13: 3–4.
50. Tagne, R.S., Telefo, B.P., Nyemb, J.N., Yemele, D.M., Njina, S.N., Goka, S.M.C., Lienou, L.L., Kamdje, A.H.N., Moundipa, P.F. and Farooq, A.D., 2014. Anticancer and antioxidant activities of methanol extracts and fractions of some Cameroonian medicinal plants. *Asian Pac J Trop Med* 7: S442–S447.
51. Sahu, N. and Saxena, J., 2013. Total Phenolic & total flavonoid content of Bougainvillea Glabra choisy and Calforina gold flower Extracts. *Int. J. Pharm. Tech* 5: 5581–5585.
52. Ndukwe, I.G., Amupitan, J.O., Isah, Y. and Adegoke, K.S., 2007. Phytochemical and antimicrobial screening of the crude extracts from the root, stem bark and leaves of *Vitellaria paradoxa* (GAERTN. F). *Afr J Biotec* 6(16): 1905–1909.
53. Adebowale, K.O., Nwokocha, L.M. and Agbaje, W.B., 2013. Composition of *Cissus populnea* stems. *J Food Compos Anal* 30: 41–46.
54. Spencer, K.C. and Seigler, D.S., 1984. Short Report. The cyanogenic glycoside barterin from *Barteria fistulosa* is epiletraphyllin B. *Phytochemistry* 23:23–65.
55. Dembitsky, V.M., 2008. Bioactive cyclobutene alkaloids. *J Nat Med* 62: 1–33.
56. Essokne, R., 2011. A taxonomic treatment of the genus *Adenocarpus* (*Leguminosae*). PhD Thesis, University of Reading.
57. Omeje, E.O., Osadebe, P.O., Akira, K., Amal, H., Adbessamad, D., Esimone, C.O., Nworu, C.S. and Proksch, P., 2012. Three – (-) – Catechin-*O*-Rhamnosides from the Eastern Nigeria Mistletoe with potent immunostimulatory and antioxidant activities. *Biomolecules* 1(1): 23–32.
58. Simić, A., Manojlovic, D., Šegan, D. and Todorovic, M., 2007. Electrochemical behaviour and antioxidant and prooxiant activity of natural phenolics. *Molecules* 12: 2327–2340.
59. Omeje, E.O., Osadebe, P.O., Nworu, C.S., Akira, K. and Proksch, P., 2011. Immuno-modulatory activity of a lupane triterpenoid ester isolated from the eastern Nigeria mistletoe, *Loranthus micranthus* (Linn). *Asian Pac J Trop Med* 11: 514–522.
60. Omeje, E.O., Osadebe, P.O., Nworu, C.S., Amal, H., Adbessamad, D., Esimone, C.O., Akira, K. and Proksch, P., 2011. Steroids and triterpenoids from Eastern Nigeria Mistletoes, *Loranthus micranthus* Linn. (*Loranthaceae*) parasitic on *Kola acuminata* with immunomodulatory Potentials. *Phytochem Lett* 4: 357–362.
61. Omeje, E.O., Osadebe, P.O., Nworu, C.S., Nwodo, J.C., Obonga, W.O., Akira, K., Esimone, C.O. and Proksch, P., 2011. A novel sesquiterpene acid and an alkaloid from leaves of the Eastern Nigeria mistletoe, *Loranthus micranthus* with potent immuno-stimulatory activity on C57BL6 mice splenocytes and CD69 molecules. *Pharm Biol* 49(12): 1271–1276.
62. Osadebe, P.O. and Omeje, E.O., 2009. Comparative acute toxicity and immunomodulatory potentials of five Eastern Nigeria mistletoes. *J. Ethnopharmacol* 126(2): 287–293.

5 Chemometric Evaluation of Phytochemical Principles in Cameroonian Plants

Ezealisiji Kenneth Maduabuchi
University of Port Harcourt

CONTENTS

DOI: 10.1201/9780429506734-5

5.1 OVERVIEW ON MEDICINAL PLANTS

Plants have not only been used in Cameroon but all around the world as safe natural sources of food, shelter, biofuel, and traditional medicines [1]. The curative potential of this phytochemical principle plays an important role in the lives of the indigenes as they rely on herbals for therapeutic ailments. The science of traditional medicine was passed down verbally through generations in the form of folklore tales. In modern civilization, there is a great need for proper documentation and pharmacological examination of these beneficial phytochemical principles. Herbal medicines entail plant parts or whole plant including the fruiting top, which can provide medicinal or curative benefit to mankind [2]. In recent times, there are only few plant species in Cameroon, which have been properly screened or phytochemically profiled and some with no proper documentation. Thus, more research is still needed to explore and profile the beneficial phytocompounds from Cameroonian plants. The pharmacologically active principles from plants are important sources of 'leads' for newer drug discovery, synthesis, and design. Subsequently, the demand for herbal formulations in various forms including herbal teas and antioxidant remedies is gaining more attention in the developing world including Cameroon. Modern hyphenated analytical equipment and software techniques are now available and useful for extraction, identification, and structure elucidation of a considerably large array of secondary plant metabolites, which constitute a source of bioactive principles [3]. Hence, a thorough screening of phytochemicals and chemometric profiling has emerged as new areas of science with the aim of standardization of natural products of plant origin. The use of Fourier-transform infrared spectroscopy, gas chromatography, and liquid chromatography coupled with mass spectrometry (GC/MS) has been of great importance in the analysis of phytochemical principles in medicinal plants [4].

5.2 SOME CRITICISM AGAINST TRADITIONAL HERBAL MEDICINES

One of the criticisms against the use of traditional medicines is that the herbal preparations are not standardized, and crude extracts or decoctions are administered in glass cup measures and not as an accurate dose compared with what is obtainable in orthodox medicine where tablets, suspensions, syrups, and injectables are prescribed in quantifiable doses, to be taken at a stated hourly interval [5]. For meaningful standardization, the knowledge of the phytochemical principle present in the plant parts is very essential. Chemistry of medicinal agent and the percent composition are important for effective use as medicinal herbs in the society. In all ramifications, the inventory and register of medicinal plants in Cameroon and their chemometric profile will be of interest to the World Health Organization who has in recent years advocated for the inclusion of traditional medicines into the hospital health management plans especially for the effective treatment of prevailing disease condition in the developing countries. Articulation of the true chemotypes present in traditional herbal medicines of Cameroon provides substantial scientific evidence in support of the efficacy of certain remedies and justifies the ethnomedicinal uses of the acclaimed medicinal plant part or fruiting top. Intensive research in phytochemistry is very important not only involving peripheral phytochemical search but a painstaking search to establish the time of the day, season, night, or morning when a particular phytochemical principle will be prevalent in the medicinal plant. For example, the dried fallen leaves of *Terminalia catappa* (umbrella tree) are preferred by the natives of Cameroon because they consider them more active than the green leaves in the management of infertility and malarial fever. Chemometric profiling shows that the dried leaves of *Terminalia catappa* are rich in phenolates and antioxidants; this justifies its ethnomedicinal uses [6].

5.3 SOME MODERN PHYTOCHEMICAL ANALYTICAL APPROACHES

5.3.1 THIN LAYER CHROMATOGRAPHY AND OTHER BIO-AUTOGRAPHIC APPROACHES

Thin layer chromatography is one of the most important techniques used in phytochemical screening: this method is simple, fast, and cost effective. It gives the researcher a quick insight as to the number of the expected phytochemical components in a mixture. Identification using ultra-violet light or chemical staining are the most prevalent techniques. Phytochemical compounds with antimicrobial activity can be evaluated using bio-autography [7].

5.3.2 HIGH-PERFORMANCE LIQUID CHROMATOGRAPHY

High-performance liquid chromatography (HPLC) is indispensable for chemometric profiling of natural or phytochemical products. It is a versatile, specific, and robust phytochemical screening technique widely used for isolation of phyto-active medicinal agents and offers 'fingerprints.' Extension to hyphenated techniques such as GC-MS or HPLC-MS is now available [8–10]. Use of diode array detectors allows each phytochemical compound to present a characteristic peak at a particular wavelength.

5.4 PHYTOCHEMICAL SCREENING USING LABORATORY REAGENTS

Phytochemical investigation through screening is easy and economical and may not require cumbersome equipment. Chromophoric reagents are used in this case. It is based on the principle of developing different colors when these reagents come into with the phytochemical principles. Precipitates and color changes and at times production of gas with different smells often confirm the presence of phyto-active agent. Results are tabulated based on organoleptic observations.

5.5 SPECTROSCOPIC PHYTO-CHEMOMETRIC ANALYTICAL TOOLS

5.5.1 ULTRAVIOLET-VISIBLE SPECTROSCOPY

Ultraviolet-visible spectroscopy deals with spectral absorption in the ultraviolet-visible region. The principle behind ultraviolet-visible spectroscopy mainly resides in the use of electromagnetic radiation between 190 and 800 nm. Two salient regions include the ultraviolet (190–400 nm) and visible (400–800 nm). For example, the quantification of alkaloids in a plant extract is achieved using an ultraviolet spectroscopic assay. The UV absorption peaks in the region of 214–486 nm are characteristic for flavonoids and their congeners [11]; they may present as a single peak or two absorption maxima in the ranges 228–286 nm (band I) and 312–360 nm (band II).

5.5.2 FOURIER-TRANSFORM INFRARED (FTIR) SPECTROSCOPY

The infrared spectral analysis offers the most outstanding number of characteristic properties of compounds. It is known to provide powerful analytical data for extensive and concise study of molecular structures of phytocompounds. This characteristic absorption corresponds to the bonds present in the molecules. Two prominent regions in the infrared spectrum are normally observed namely: (i) group frequency region (4000–1300 cm^{-1}) and (ii) the fingerprint region (300–400 cm^{-1}). Below are some phytochemicals and their corresponding IR absorption bands.

Phytochemicals	IR absorption bands (cm^{-1})
[i] Polyphenols	2500–3500
[ii] Amides	3100–3500
[iii] Glycosides	650–2000
[iv] Antioxidants	2400–3600
[v] Tannins	3000–3700
[vi] Alcohols	3200–3550

5.5.3 MASS SPECTROMETRY

Modern gas chromatography and liquid chromatography are now equipped with mass spectrometers, which serve as the detector. Molecular fragments can be picked up and reassembled for proper identification of phytochemical compound present in the plant extract.

5.6 PHYTOCHEMISTRY OF *TETRACARPIDIUM CONOPHORUM* (CAMEROONIAN WALNUT)

Taxonomy [12]

The taxonomy of the Cameroonian walnut is represented below;

Kingdom	Plantae	Plant
Subkingdom	Tracheobionta	Vascular plant
Super division	Spermatophyta	Seed plant
Division	Magnoliopsida	Flowering plant
Class	Magnoliopsida	Dicotyledons
Sub-class	Rosidae	-
Order	Euphorbiales	-
Family	Euphorbiaceae	Spurge family
Genus	*Tetracapidium*	Pax
Species	*Conophorum*	-
Scientific name:	Tetracarpidium	Conophorum (Mull.arg.) Hutch&Daizel
Bakassi name (local):	Ngak	

5.6.1 WALNUTS [13]

Walnuts are eaten as snacks in general and they are seeds of trees of the genus Juglans, which include the English (Persian) walnut. The species (*J. regia*) is indigenous to Persia, while the black walnut (*J. nigra*) is native to Eastern America. Walnut seeds are rich in nutritional values and contain essential fatty acids as well as proteins. The walnut fruit is enclosed in a green, leathery, fleshy husk which is inedible. After harvest, the removal of the husk reveals the 'winky' walnut shell which is in two halves. The shell is hard and encloses the kernel which is made up of two halves separated by partition. The seeds are rich in antioxidants and contain the walnut oil. The husk of the black walnut *J. nigra* is used to make an ink for writing. They are also used as organic dye for fabrics due to their high phenolate content.

5.6.2 CAMEROONIAN WALNUT [14]

5.6.2.1 Description

Tetracarpidium conophorum (Mull Arg. Conophr.) Euphorbiaceae commonly known as the Cameroonian walnut is a West Equatorial perennial climbing shrub 10–20 feet long, found growing wild in the mist forest zone of sub-Saharan Africa. In Cameroon, the plant can be found in the Bakassi Peninsular and the western Cameroon at large. The indigenes of Cameroon call it Ngak. The Cameroonian walnut has been useful ethno-medically among the indigenes in several complementary medicines to treat numerous health challenges such as diabetes mellitus, gastro-intestinal ulcer, anti-depressants, analgesic, hyperlipidemia, male infertility, uterine fibroid, and wound healing as well as an antimicrobial agent. The above ethnomedicinal claims on

TABLE 5.1
Phytochemical Constituents Present in the Nuts of *Tetracarpidium Conophorum* [14]

S/No	Constituents	Crude Kernel	Methanol Extract
	Saponin	+	+
	Alkaloid	+	+
	Glycosides	-	-
	Carbohydrate	+	+
	Fixed oil	+++	-
	Tannin	+++	+
	Flavonoids	++	+++
	Terpenoids	+++	+++
	Steroid	+	+
	Triterpenes	+	+
	Proteins	++	++

+, slightly present; ++, moderately present; +++, abundantly present; -, absence.

TABLE 5.2
UV-VIS Values of the Different Nut Extracts of *Tetracarpidium Conophorum* [14]

Methanol		N-Hexane	
Nm	Absorbance	Nm	Absorbance
415	2.230	482.80	2.159

TABLE 5.3
FTIR Peak Values (cm⁻¹) and Functional Groups of Different Phyto-Components Identified in the Methanol Nut Extract of *Tetracarpidium Conophorum* [14]

Peak Values	Identity	Description	Likely Functional Group Present	Inference (nm)
2930	14.728	Sharp and strong	Aliphatic C-H stretching	C-H
2671	54.348	Broad and weak	Saturated C-H stretching	C-H
1720	25.667	Sharp and strong	C=O(carbonyl, aldehyde, ketone, carboxylic acids ester)	Present
1548	59.833	Medium-weak multiple band	C=C stretching	C=C present
1438	40.036	Sharp and variable	-C-Stretching	-CH-
1258	42.083	Variable	Acidic C-O stretching	C-O
929	56.719	Weak	=C-H bending	
714	51.076		Aromatic =C-H mono substitution	

TABLE 5.4
FTIR Peak Values (cm⁻) and Functional Groups of Different Phyto-Components Identified in the N-Hexane Extract of the Nuts of *Tetracarpidium Conophorum* [14]

Peak Values	Identity	Description	Likely Functional Group Present
3444	10.26	Strong band	O-H of alcoholic or phenol that is H- bonded.
2942	82.106		Saturated C-H stretching.
1638	45.508	Variable	Conjugated double bond, C=C stretching
1402	85.489	Variable	C=H in alkane bending mode.
1245	87.975	Very weak	C-O stretching of ester, carboxylic acid.
1050	88.095	Very weak	C-O stretching in ether, alcohols and esters.

TABLE 5.5
Phytocomponents identified in the methanolic extract of *Tetracarpidium conophorum* by GC-MS [14]

S/No	Retention Time (min)	Name of Compounds	Molecular Formulas	Molecular Weight (gmol⁻¹)	Peak Area (%)
	4.561	Dimethyl butanedioate	$C_6H_{10}O_4$	146	0.46
	5.144	3,5-Octadiene-2-one	$C_8H_{12}O$	124	0.77
	5.851	Methyl caprylate	$C_9H_{18}O_2$	158	0.34
	8.069	4H-1,2,3-Triazole-4-carboxylic acid, 5-phenyl-1-oxide	C_9H_7O3	205	1.57
	8.367	1H-Indole	C_8H_7	117	1.05
	10.052	Methyl azelaaldehydate	$C_{10}H_{18}O_3$	186	3.20
	10.809	Diosphenol	$C_{10}H_{16}O_2$	168	1.95
	11.475	Dimethyl nonanedioate	$C_{11}H_{20}O_4$	216	5.52
	16.815	Methyl tridecanoate	$C_{14}H_{28}O2$	228	16.14
	19.732	2-Chloroathyl linoleate	$C_{20}H_{35}CIO_2$	342	6.41
	21.412	Cis, cis-linoleic acid	$C_{18}H_{32}O_2$	280	11.46
	23.339	Methyl (8E,11E,14E)-8, 11,14-icosatrienoate	$C_{21}H_{36}O_2$	320	31.61
	24.465	Bis (2-dimethylamino) ethyl] ether	$C_8H_{20}N_2O$	160	4.03
	24.859	Methyl docosanoate	$C_{23}H_{46}O_2$	354	5.17
	25.046	n-Octyl phthalate	$C_{24}H_{36}O_4$	390	2.32
	25.593	Methyl heptacosanoate	$C_{28}H_{56}O_2$	424	1.23
	26.679	Methyl melissate	$C_{31}H_{62}O_2$	466	1.04
	27.873	Squalene	$C_{30}H_{50}$	410	5.73

R/T, Retention time.

TABLE 5.6
Phyto-Components Identified in the n-Hexane Nut Extract of *Tetracarpidium Conophorum* by GC-MS [14]

S/No	Retention Time (min)	Name of Compound	Molecular Formula	Molecular Weight (gmol⁻¹)	Peak Area (%)
	3.496	Dimethyl butanedioate	$C_6H_{10}O$	98	0.16
	6.171	3,5-Octadiene-2-one	$C_7H_7O_2$	137	2.19
	6.884	Methyl caprylate	$C_3H_8O_3$	92	8.85
	7.379	4H-1,2,3-Triazole-4-carboxylic acid, 5-phenyl-1-oxide	$C_6H_8O_4$	144	0.18
	10.883	1H-Indole	$C_{10}H_{16}O_2$	168	0.83
	12.765	Methyl azelaaldehydate	$C_{19}H_{36}O_2$	296	0.60
	16.870	Diosphenol	$C_{14}H_{28}O_2$	228	0.16
	18.677	Dimethyl nonanedioate	$C_{16}H_{32}O_2$	256	6.08
	20.057	Methyl tridecanoate	$C_{19}H_{32}O_2$	296	9.57
	21.050	2-Chloroathyl linoleate	$C_{18}H_{34}O_2$	282	57.27
	23.835	Cis, cis-linoleic acid	$C_{36}H_{66}O_2$	546	3.75
	24.191	Methyl (8E,11E,14E)-8,11, 14-icosatrienoate	$C_{16}H_{18}N_2O_3$	286	3.84
	26.395	Bis (2-dimethylamino) ethyl] ether	$C_{21}H_{40}O_4$	356	6.52

Cameroonian walnut has been justified as revealed by the various findings shown in Tables 5.1–5.6 which gave different types of phytochemical constituents, peak values, functional groups in different phytochemical components as identified by phytochemical screening, FTIR, GC-MS, and UV spectroscopy, respectively. Fixed oils, tannins, and terpenoids were present in abundance in the crude extracts of Tetracarpidium conophorum kernel extract. The methanol extracts contain abundant yield of flavonoids and terpenoids, and the alkaloid content was minimal as both extracts contain no glycoside as shown in Table 5.1. The maximum wavelength of absorption for the methanol and N-hexane nut extract of *Tetracarpidium conophorum* was observed at 415.00 and 482.80 nm, respectively, with an absorbance peak of 2.230 and 2.159 (see Table 5.2). The methanol extract gave different FT.IR peak values at 2929.97 and 2671.50; these peak characteristics were shown to be sharp, strong, and likely to be functional groups associated with aliphatic C-H stretching vibration. Other peaks were observed to conform to those of C=C stretching and C=O stretching vibration which are common to ethers or alcohols as shown in Table 5.3. Table 5.4 clearly shows the stretching characteristic peaks in ether, alcohols, and esters; the strong O-H band and very week C-O stretch depicts alcohol and carboxylic acid, respectively. Pharmacological active principles were found in varying proportions, ranging from essential fatty acids to indole moiety as well as squalene in methanol extracts, as presented in Table 5.5 and N-hexane extracts in Table 5.6. The different nut extracts of *Tetracarpidium conophorum* as shown in Table 5.7 showed a remarkable biological

TABLE 5.7

Biological Activities of Some Active Principles Present in the Different Extracts of *Tetracarpidium Conophorum nut* [14]

Phytocomponent	Nature and Compounds	Biological Activities[a]
2, 12-Octadecadienoic acid (Z,Z)-Cis, cis-linoleic acid. (Linoleic)	Unsaturated fatty acid	5-Alpha reductase inhibitor, antiandrogenic, antiarteriosclerotic, anticoronary, antifibrinolytic, antihistaminic, antiinflammatory, antileukotrienic, antiprostatic, hepatoprotective, carcinogenic, immunomodulator, antieczemic, hypocholesteramine
1-H Indole	Indole	Aldose-reductase-inhibitor, antiacne, antibacterial, anticariogenic, antisaimonella, antiseptic, antistreptococcic, cancer preventive, carcinogeniclnsectiphile, nematicide, perfumery, fragrance, pesticide, tumorigenic.
Squalene	Triterpene	Antibacterial, antioxidant, antitumor, cancer-preventive, chemo-preventive, immunostimulant, lipoxygenase inhibitor, perfumery, pesticide, sunscreen.
9- Octadecadienoic acid (Z)- (Oleic acid)	Unsaturated fatty acid	5-Alpha-reductase-lnhibitor, allergenic, alpha-reductase-inhibitor, anemiagenic, antialopecic, antiandrogenic, antiinflammatory, antileukotriene, cancer-preventive, choleretic, dermatitigenic, flavor, hypoeholesterolemic, insectifuge, irritant, percutaneostimulant, perfumery propecic;
N-Hexadecanoic acid. (Palmitic acid)	Saturated fatty acid	Antioxidant, hypocholesterolemic, nematicide, pesticide, lubricant, antiandrogenic, flavor, hemolytic, 5-alpha reductase inhibitor
Methyl (8E,11E,14E)-8,11, 14-eicosatrienoate	Unsaturated fatty acid ester	Antiarthritic, anticoronary, antiinflammatory
Tridecanoic acid methyl ester	Unsaturated fatty acid esters	Antioxidant, perfumery.
Glycerol	Fatty alcohol	Anticataract, antiearwax, antiketotic, antineuralgic, arrhythmigenic, emollient, hyperglycemic
Methyl nicotinate	Nicotinic acid methyl ester	AntiCrohn's, antidote, (pesticides), antischizophrenic, antithyrotoxic, choleretic, hypoeholesterolemic, hypoglycemic, insulinase inhibitor, insufinotonic, lipolytic

Source: Dr. Duke's Phytochemical and Ethnobotanical Databases [Online database].

[a]   could also be referred to as pharmacological activities.

activity ranging from antiartritic, anticarcinogenic, antiartriosclerotic, and antibacterial to antifibrinolytic potential. All these were in agreement with Dr. Duke's phytochemical and ethnobotanical database {online database} [15].

5.7 PHYTOCHEMISTRY OF *ANNONA MURICATA*

Taxonomy [16]

The taxonomy of *Annona muricata* is represented below

Domain	Eukaryota
Kingdom	Plantae
Phylum	Spermatophyta
Subphylum	Angiospermae
Class	Dicotyledonae
Order	Annonales
Family	Annonaceae
Genus	*Annona*
Species	*Muricata*
Botanical name	*Annona muricata* L.
Synonyms	*Annona crassiflora* Mart, *Guanabanus muricata* L
Plant type	Perennial, seed propagated (tree).

5.7.1 ANNONA MURICATA L.

Annona muricata belongs to the family Annonaceae, a fruit tree with a long history of ethnomedicinal use in the Cameroons. Common names include soursop, graviola, or guanabana. The fruiting top and roots are responsible for a wide array of medicinal activities, and indigenous Cameroonians use this plant in their folk medicines. Phytochemical investigation has justified the medicinal values including anticancer, anticonvulsant, antimalarial, hepatoprotective, and anti-diabetics. Studies have revealed that annonaceous acetogenins are the major constituent of *A. muricata*.

5.7.1.1 Description

Annona muricata L. belongs to the Annonaceae family with close to 125 genera and 2400 species. It is native to the warmest tropical areas in the Americas and now widely distributed in other areas such as India, Malaysia, Nigeria, and Cameroon. The soursop is a free branching tree which could reach a height of 26 or 32 feet. The leaves are evergreen and odorous due to the presence of essential oils. The flowers are singly borne and may emerge anywhere on the trunk or branches. The fruits are heart-shaped as shown in Figure 5.1, oval, or curved with varying sizes ranging from 3 to 10-inch long and up to 5-inch width. They are greenish yellow, succulent, and soft to touch when ripe.

5.7.2 ETHNOMEDICINAL USES

The fruit serves as a natural medicine for arthritis, dysentery, diarrhoea, neuralgia, fever, malaria, rheumatism, skin rashes, and antihelmintics, while the leaves and

FIGURE 5.1 Green fruit of *A. muricata*. (Courtesy of Ezealisiji K.M)

rootbark are believed to treat abscesses and rheumatism. In South America and tropical Africa, including Nigeria and Cameroon, the leaves of *A. muricata* are used against tumors and cancer [17].

5.7.3 PHYTOCHEMISTRY

Phytochemical search on different parts of *A. muricata* plant have revealed the presence of different phytoconstituents and including alkaloids, megastigmanes, flavonols, triglycosides, phenolics, cyclopeptides, and essential oils. The species *A. muricata* is rich in annonaceous acetogenin compounds (AGEs) [18]. Some biological activities of various phytochemical principles present in *Annona muricata* are presented in Table 5.8, and turpinoside A, kaempferl-3, rutin, citroside A, and cohibin A were implicated in wound healing, antihypertensive, antimicrobial, and anticancer activities.

TABLE 5.8
Biological Activities of Some Phytochemical Principles Present in *A. Muricata*

Phytochemical Compounds	Nature of Compound	Biological Activities
Turpinoside A	Glycoside	Wound healing
Kaempferol 3-*O*-rutinoside	Polyphenol	Antihypertensive
Kaempferol 3-*O*-robinobioside	Polyphenol	Antihypertensive
Rutin	Flavonoid	Anticancer, antioxidant
Blumenol C	Sesquiterpenes	Antimicrobial
Citroside A	Terpene glycoside	Antioxidant
Cohibin B	Fatty acid-lactone	Immunosuppressive
Cohibin A	Fatty acid-lactone	Antimicrobials

5.8 PHYTOCHEMISTRY OF *TERMINALIA CATAPPA*

Taxonomy [19]

Kingdom	Plantae	Plants, Planta, Vegetal plants
Sub-kingdom	Viridiplantae	-
Infra-kingdom	Streptophyta	Land plants
Super-division	Embryophyta	-
Division	Tracheophyta	Vascular plants, tracheophytes
Sub-division	Spermatophyta	Spermatophytes, seed plants.
Class	Magnoliopsida	-
Superorder	Rosanae	-
Order	Myrtales	-
Family	Combretaceae	Combretums
Genus	*Terminalia* L	Tropical almond
Species	*Terminalia catappa*	Indian almond, tropical almond.

5.8.1 DESCRIPTION

Terminalia catappa (Combretaceae) is commonly referred to as Indian almond or tropical almond. It is a leafy tree that grows in the tropical parts of the world including Brazil, East and central Africa, the Caribbean, and India. A spreading fibrous root is a characteristic which makes it stable in the coastal regions and tolerant to strong winds. The tree grows to a height of 24–45 m forming a large crown spread in an open umbrella-like manner as shown in Figures 5.2 below. The flowers are small, white or cream in color with a mildly unagreeable smell. The leaves present as dark-green but turn yellow with time and then red before dropping off. The fruit is ovate in

FIGURE 5.2 Fruiting top of *T. catappa*. (Courtesy of Ezealisiji K.M)

shape, drupe, and the color may change from green, yellow, and then wine-red when it is fully matured. The fruit contains a seed that houses the kernel, rich in edible oil.

5.8.2 ETHNOMEDICINAL USES OF *TERMINALIA CATAPPA*

Terminalia catappa has been used by man as a source of traditional medicine since ancient time. The fruits' leaves and root bark are used for treating dysentery in Asia, while in South America, the young leaves are used to treat colic and abdominal pains. In Nigeria and Central Cameroon, the indigenes pick up the fallen yellow-ish-red leaves, which they boil and then drink the hot aqueous extract to improve fertility in women. The aqueous extract of *Terminalia catappa* is also used as an anti-malaria agent in the above African localities. The bark and root have also been used in fever, dysentery, diarrhea, and oral thrush. The people of Tonga and Samoa have used this plant for treating mouth infection by boiling and drinking the decoction of the young leaves. The ethnomedicinal claims above have been strongly supported by the findings in Tables 5.9–5.13, which gave different types of secondary metabolites, peak values, and corresponding functional groups of the different phytochemical compounds and names of phytochemical principles as identified by phytochemical screening, FTIR, and GC-MS spectroscopy respectively, the secondary metabolites present in various extracts of the leaves of *Taminalia catappa* are shown in Table 5.9 with Tannin and phlobatannin as well as flavonoids occurring in abundance with few saponins and anthraquinone [20]. Table 5.10 shows that N-hexane leaf extract of *Taminalia catappa* gave various peaks at 2922.25, 2852.53, and 1453.79 cm^{-1}, peak characteristics were sharp, strong, and intermediate, and these are likely to be aliphatic stretching vibrations of terminal C-H of an alkyl or alkene functional group, while the dichloromethane extract as shown in Table 5.11 gave outstanding peaks at 3393.67, 2958.36, and 2850.57 corresponding to broad, sharp, and strong bands for O-H stretch of alcohols and C-H stretch of alkenes and alkanes, respectively. The dichloromethane leaf extract of *Taminalia catappa* was found to be rich

TABLE 5.9
Secondary Metabolites Present in Various Extracts of the Leaves of *T. Catappa* [20]

Class of Compound	N-Hexane	Dichloromethane	Methanol
Tannins	-	+	+++
Phlobatannins	-	-	++
Saponins	+	-	+
Carbohydrates	+	++	++
Flavonoid	-	-	++
Alkaloid	++	-	-
Anthraquinone	-	-	-
Reducing sugar	-	-	++

+, Present; -, absent; ++, moderately present; +++, abundantly present.

TABLE 5.10
FTIR Peak Values and Corresponding Functional Groups of the Different Phyto-Chemical Compounds Identified in the N-Hexane Extract of the Leaves of *Terminalia Catappa*. [20]

Peak Values (cm⁻¹)	Description	Likely Group Present	Functional Group Inference
2922	Sharp and strong	Aliphatic stretching	C-H C-H
2853	Sharp, intermediate	Terminal C-H of an alkyl, alkene	C-H
1454	Sharp, intermediate	C-H bending	C-H
1243	Very weak	C-O stretching in either, alcohols, and esters	C-O
836	Very weak and sharp	=C-H bending	=C-H
722	Weak, sharp	CH long chain methyl rocking	C-H

TABLE 5.11
FTIR Peak Values and Corresponding Functional Groups of the Different Phytochemicals Identified in the Dichloromethane Extracts of the Leaves of *Terminalia Catappa* [20]

Peak Values (cm⁻¹)	Description	Likely Group Present	Functional Group Inference
3394	Broad and strong	O-H of an alcohol or phenol	O-H
2958	Strong and sharp	Terminal C-H of an alkyl, C-H alkene	C-H
2851	Sharp, variable	C-H stretching of alkane, alkene	C-H
1659	Weak	C=C stretching	C-C
1164	Very weak	C-O stretching in an alcohols, ethers	C-O

TABLE 5.12
Phytochemical Compounds Identified in the Dichloromethane Extract of the Leaves of *Terminalia Catappa* by [GC-MS] [20]

S/N	Retention Time (min)	Name of Compounds	Molecular Formula	Molecular Weight (gmol⁻¹)	Peak Area (%)
	16.980	n-Hexadecanoic acid	$C_{16}H_{32}O_2$	256	35.18
	18.354	6-Chlorododecanoic acid, chloromethyl ester	$C_{13}H_{24}ClO_2$	282	20.36
	18.511	Octadecanoic acid	$C_{18}H_{36}O_2$	284	8.32
	21.304	Glycerol-1-palmitate	$C_{19}H_{38}O_4$	330	18.51
	21.573	Chloroacetic acid	$C_{18}H_{35}ClO_2$	318	17.63

TABLE 5.13

Phytochemical Compounds Present in the N-Hexane Extract of the Leaves of *Terminalia Catappa* by [GC-MS] [20]

S/N	Retention Time (min)	Name of Phytocompound	Molecular Formula	Molecular Weight (gmol⁻¹)	Peak Area (%)
	10.079	1,8-Nomadien-3-0l	$C_9H_{16}O$	140	0.73
	10.934	Alpha bergamotene	$C_{15}H_{24}$	204	0.96
	11.650	Chloroacetic acid octyl ester	$C_{10}H_{19}ClO_2$	206	2.52
	12.271	1,3;2,5-Diformal-1-rhamnitol acetate	$C_{10}H_{12}O_8$	260	0.24
	13.260	Decanynoic acid	$C_{10}H_{16}O_2$	168	0.41
	13.690	Piperidine	$C_7H_{12}ClNO$	161	0.28
	14.034	Tetrahydrofuran-2-one-3-[1 fluoroethyl]-5{[2]hydroxypropyl] benze-ethyl	$C_{17}H_{23}FO_3$	294	1.45
	14.227	Phthalofyne	$C_{14}H_{14}O_4$	246	0.78
	14.647	Benzenesulfonamide	$C_6H_7NO_2S$	157	1.60
	15.264	Tetradecanoic acid (myristic acid)	$C_{14}H_{28}O_2$	228	2.25
	15.869	Chloroacetic acid, decyl ester	$C_{12}H_{23}ClO_2$	234	6.95
	16.150	n-Pentadecanoic acid	$C_{15}H_{30}O_2$	242	1.02
	16.850	cic-9-Hexadecanoic (palmitoleic) acid	$C_{16}H_{30}O_2$	254	3.66
	17.009	n-Pentadecanoic acid (Palmitic acid)	$C_{16}H_{32}O_2$	256	35.63
	17.530	Chloroacetic acid, 4-pentadecyl ester	$C_{17}H_{33}ClO_2$	304	0.52
	17.773	n-Heptadecanoic (margaric acid)	$C_{17}H_{34}O_2$	270	1.45
	18.377	Oleic acid	$C_{18}H_{34}O_2$	282	21.14
	18.528	Stearic acid	$C_{18}H_{36}O_2$	284	11.44
	20.441	1,2-Dipalmitin	$C_{35}H_{68}O_5$	568	3.18
	20.647	Chloroacetic acid, 3-tridecyl ester	$C_{15}H_{29}ClO_2$	276	0.97
	20.840	Adipic acid, beta-citronellyl heptyl ester	$C_{23}H_{42}O_4$	384	1.88
	20.988	Cyclohexanecarboxylic acid	$C_{18}H_{32}O_2$	280	0.55

in fatty acids and glycerol as shown in Table 5.12, while the N-hexane extracts in Table 5.13 contain piperidine and sulfonamides with fatty acids. Table 5.14 shows the biological activities of some active phytocompounds in various leaf extracts of T. catappa. These include hypercholesterolemic, antialopecia, antioxidant, and anthelmintic activities especially for whipworms.

TABLE 5.14

Biological Activities of Some Active Principles found in the N-Hexane Extracts of the Leaves of *T. Catappa* [20]

Phytocomponents	Nature of Compounds	Biological Activities[a]
n-Hexadecanoic acid	Saturated fatty acid	Hypercholesterolemic, hemolytic, 5-alpha-reductase-inhibitor, antialopecia, antioxidant, antifibrinolytic, lubricant, nematicide
Phhalic acid, mono (1-ethyl-1 methyl-2 propynyl) ester (pthalofyne)	Aromatic dicarboxylic ester	Anthelminthic especially for whipworm (Pubchem database)
Tetradecanoic acid (myristic acid)	Saturated fatty acid	Cancer-protective, cosmetic, nematicide, lubricant, hypercholesterolemic
n-Pentadecanoic acid	Saturated fatty acid	Antioxidant, lubricant,
Cis-9hexadecanoic acid (palmitoleic acid)	Unsaturated fatty acid	Increases insulin sensitivity and inhibits destruction of insulin secretory pancreatic beta cells (Yang et al., 2011)
Heptadecanoic acid (margaric acid)	Saturated fatty acid	Antioxidant, it can be used to detect elevated insulin with associated high ferritin as well as resolve metabolic syndrome and decrease high ferritin levels (Venn-Watson et al., 2015)
Cis-9 Octadecanoic acid (oleic acid)	Unsaturated fatty acid	5-alpha reductase inhibitor, allergenic, cancer preventive, alpha reductase inhibitor, percutostimulant, dermatitigenic, antialopecic, propecic, hypocholesterolemic, perfumery, anti-inflammatory
n-Octadecanoic acid (stearic acid)	Saturated fatty acid	Propecic, hypocholesterolemic, lubricant, flavor, cosmetics, 5-alphareductase inhibitor, suppository

Source: Dr. Duke's Phytochemical and Ethnobotanical Database {Online Database}, unless stated otherwise in table.

[a] Could also be referred to as pharmacological activities.

5.8.3 Phytochemistry of *Cnidoscolus Aconitifolius* [21]

5.8.3.1 Taxonomy

Botanical Description	
Super-kingdom	Eukaryote
Kingdom	Plantae
Sub-kingdom	Tracheobionta
Super-division	Spermatophyta
Division	Magnoliophyta
Class	Magnoliopsida
Sub-class	Rosidae
Order	Malpighiales
Family	Euphorbiaceae
Sub-family	Crotonideae
Genus	*Cnidoscolus*
Species	*Aconitifolius*
Synonyms	*Cnidoscolus chayamansa, Jatropha aconitifolia*

5.8.3.2 Description

Cnidoscolus aconitifolius (Euphorbiaceae) is also referred to as cabbage star, 'Efo iyana ipaja' in Nigeria. In the central areas of Cameroon, *Cnidoscolus aconitifolius* is a free-growing, ornamental, evergreen, and deciduous shrub that attains the height of up to 7 m with starlike arranged palmate lobular leaves. The leaves contain some milky sap, and the flowers are small. The plants grow freely in Central America and Central Africa and are abundant in the tropical areas of Nigeria and Cameroon. They prefer growing wild in humid environments. *Cnidoscolus aconitifolius* is rich in protein and natural antioxidants, and it is used locally to treat skin diseases.

5.8.4 Ethnomedicinal Uses

The fruiting top is used to manage arteriosclerosis and related conditions. *C. aconitifolius* is used as a laxative, improving blood circulation and for good vision, among the early Cameroonians and the neighboring countries of West Africa. It also serves as an anti-inflammatory agent for veins and hemorrhoids and is used to treat scorpion stings. The white milky sap when applied to pimples and warts produces good healing results. It also cures skin blemishes [22]. All the above ethnomedicinal claims are well justified by results obtained from various spectroscopic techniques as shown in Tables 5.15–5.19. The FT-IR analysis of *Cnidoscolus aconitifolius* revealed the presence of alkaloids due to N-H stretching, flavonoids, and polyphenols as shown in Table 5.15. The O-H stretching of polyphenols was outstandingly present. The phytocomponents in the N-hexane extracts of *Cnidoscolus aconitifolius* showed abundant presence of fatty acids as shown in Table 5.16, while the dichloromethane extracts revealed 9-eicosyn

TABLE 5.15

FTIR Peak Values (cm^{-1}) and Functional Groups of Different Phyto-Components Extracted from the Leaf of *Cnidoscolus Aconitifolius* [21]

N-Hexane		Dichloromethane		Ethyl Acetate		Methanol	
Peak values	Functional group	Peak values	Functional group	Peak values	Functional group	Peak values	Functional group
3460	O-H stretch, H-bonded	3419.90	O-H stretch, H-bonded	3525.99	O-H stretch, H-bonded	3471.98	O-H stretch, H-bonded
2916	CH$_2$ stretch symmetric	2920.32	CH$_2$ stretch asymmetric	2939.61	CH$_2$ stretch asymmetric	2929.97	CH$_2$ stretch asymmetric
2849	CH$_2$ stretch symmetric CSC	2850.88	CH$_2$ stretch symmetric	2088.98	CEC stretch	2131.41	CEC stretch
1647	C=C stretching nonconjugated	1705.13	C=O stretch	1722.49	C=O	1660.77	C=C stretch non conjugated
1456	CH$_3$ deformation	1464.02	CH$_3$ deformation	1633.76	N-H deformation	1417.73	CH$_3$-N {C-H deformation)
1375	CH$_3$-c- deformation	1375.29	CH$_3$-c- deformation	1446.66	C-H deformation	1303.92	C-H wag {- CH$_2$X)
1244	C-N stretch	1165.04	C-H wag (- CH$_2$X)	1384.94	C-H deformation	1076.32	C-N stretch
1036	C-N stretch	979.87	=C-H bend	1247.99	C-N Aliphatic amines	441.71	
448		443.64		1047.38	C-N Aliphatic amines	434.00	
436		430.14		981.80	=C-H bend	405.06	
422		420.50		445.57	-		
415		403.41		428.21	-		
-		-		412.78	-		
-		-		405.06	-		

TABLE 5.16
Phyto-Components Identified in the N-Hexane Extract of *Cnidoscolus Aconitifolius* by GC-MS [21]

S/No	Retention Time (min)	Name of Compound	Molecule Formula	Molecule Weight	Peak Area (%)
	17.66	6,10,14-trimethyl-2-pentadecanone	$C_{18}H_{36}O$	268	1.30
	20.69	n-Hexadecanoic acid	$C_{16}H_{32}O_2$	256	17.54
	22.66	3,7,11,15-Tetramethyl—[R-[R*,R*, R*,(E)]]-2-hexadecen-1-ol	$C_{20}H_{40}O$	296	2.55
	23.57	9-Octadecenoic acid (z)	$C_{18}H_{34}O_2$	282	56.77
	23.76	n-Octadecanoic acid	$C_{18}H_{36}O_2$	284	11.12
	26.84	1,2,3-Propanetryl ester Octadecanoic acid (E,E,E)-9-	$C_{57}H_{104}O_6$	884	3.27
	27.05	1-(hydroxymethyl)-1,2-ethanediyl Ester Hexadecanoic acid	$C_{35}H_{68}O_5$	568	1.51
	28.77	1,2,3- Propanetriyl ester octadecanoic acid (E,E,E)-9-	$C_{57}H_{104}O_6$	884	5.94

TABLE 5.17
Phyto-Components Identified in the Dichloromethane Extract of *Cnidoscolus Aconitifolius* by GC-MS [21]

S/No	Retention Time (min)	Name of Compound	Molecule Formula	Molecule Weight	Peak Area (%)
	17.44	9-Eicosyne	$C_{20}H_{38}$	278	2.20
	17.67	6,10,14-trimethyl-2-pentadecanone	$C_{18}H_{36}$	268	0.79
	17.77	3,3,6-trimethyl-4,5-heptadene-2-ol	$C_{10}H_{18}$	154	1.13
	17.96	2-(2 Hydroxypropyl)-1, 4-benzenediol	$C_9H_{12}O_3$	168	2.06
	19.17	Methyl-14-methylpentadecanoate	$C_{17}H_{34}O_2$	270	1.05
	20.69	n-Hexadecanoic acid	$C_{16}H_{32}O_2$	256	12.82
	22.34	9,129-Octadecanoic acid (z, z)- methyl ester	$C_{19}H_{36}O_2$	294	1.80
	22.40	9-Octadecenoic acid (z)-methyl ester	$C_{19}H_{34}O_2$	296	2.43
	22.67	3,7,11,15-Tetramethyl—[R-[R*,R*, R*,(E)]]-2-Hexadecen-1-ol	$C_{20}H_{40}O$	296	1.16
	23.57	9-Octacosane (z)	$C_{18}H_{34}O_2$	282	43.13
	23,76	n-Octacosanic acid	$C_{18}H_{36}O_2$	284	8.36
	25.85	n-Octacosane	$C_{28}H_{58}$	394	13.59
	26.85	2,3-Dihydroxypronpyl ester-9- octadecenoic acid	$C_{21}H_{40}O_4$	356	3.29
	28.78	1,2,3-Propanetriyl ester (E,E,E)-9- Octadecenoic acid	$C_{57}H_{104}O_4$	884	4.19

TABLE 5.18

Phyto-Components Identified in the Ethyl Acetate Extract of *Cnidoscolus Aconitifolius* by GC-MS [21]

S/No	Retention Time (min)	Name of Compound	Molecule Formula	Molecule Weight	Peak Area (%)
	6.23	1,2,3-Propanetriol	$C_3H_8O_3$	92	0.64
	7.77	1,2,3-Propanetriol monoacetate	$C_5H_{10}O_4$	134	11.18
	9.68	1,2,3-Propanetriol diacetate	$C_7H_{12}O_5$	176	5.05
	10.90	1,2,3-Propanetriol triacetate	$C_9H_{14}O_6$	218	0.43
	12.20	1,2,3-Propanetriol monoacetate	$C_5H_{10}O_4$	134	3.64
	13.02	1,2,3-Propanetriol diacetate	$C_9H_{14}O_6$	218	0.95
	13.38	1,2,3-Propanetriol monoacetate	$C_5H_{10}O_4$	134	4.50
	29.76	*l*-(+)-Ascorbic acid-2,6-dihexadecanoate	$C_{38}H_{68}O_8$	652	12.53
	23.68	9-Octadecemoic acid (z)	$C_{10}H_{34}O_2$	282	57.22
	28.79	1,2,3-Propanetriyl ester (E,E,E)-9-Octadecenoic acid	$C_{57}H_{104}O_6$	884	3.86

TABLE 5.19

Phyto-Components Identified in the Methanol Extract of *Cnidoscolus Aconitifolius* by GC-MS [21]

S/No	Retention Time (min)	Name of Compound	Molecule Formula	Molecule Weight	Peak Area (%)
	7.19	Gamma-aminobutyric lactam	C_4H_7NO	85	3.87
	12.05	DL-Poline-5-oxo-methyl ester	$C_6H_9NO_3$	143	5.80
	17.44	Tetradecyl-oxirane	$C_{16}H_{32}O$	240	1.18
	19.19	Methyl ester palmitic acid	$C_{17}H_{34}O_2$	270	2.23
	20.69	n-Hexadecanoic acid	$C_{16}H_{32}O_2$	256	11.81
	22.35	9,12-Octadecadienoic acid, (z,z) methyl ester	$C_{19}H_{34}O_2$	294	3.37
	22.41	9-Octadecenoic acid (z) methyl ester	$C_{19}H_{36}O_2$	296	5.13
	23.60	9-Octadecenoic acid (z)	$C_{18}H_{34}O_2$	282	53.97
	23.78	n-Octadecenoic acid	$C_{18}H_{36}O_2$	284	6.84
	28.79	1,2,3-Propanetriyl ester (E,E,E)-9-Octadecenoic acid	$C_{57}H_{104}O_6$	884	5.81

and different types of essential fatty acids as shown in Table 5.17. The ethyl acetate leaf extract of *Cnidoscolus aconitifolius* also contains lots of essential fatty acids and ascorbic acid, see Table 5.18, while the methanol extract of the same plant leaf contains varying quantities of fatty acids and gamma-aminobutyric lactams as shown

TABLE 5.20
Biological Activities of Some Active Principles Present in Different Extracts of *Cnidoscolus Aconitifotius* [21]

Phyto-Components	Nature of Compound	Biological Activities[a]
n-Octadecenoic acid	Fatty acid	
9-Octadecenoic acid	Fatty acid	5-α reductase inhibitor, cosmetic, flavor, hypocholesterolemic, lubricant, perfumery, propecic and suppository. Anti-inflammatory, Anti-alopecic, anemiagenic, 5-α reductase inhibitor, a-reductase inhibitor lubricant, antitumor, choleretic, dermatitigenic, immunostimulant, anti-leucotriene-D4, antiandrogenic, [b]lipoxygenase inhibitor, allergenic, flavor, hypocholesterolemic, insectifuge, irritant, percutaneostimulant, perfumery and propecic
n-Hexadecanoic	Fatty acid	Anti-alopecic, anti-androgenic, antifibrinolytic, antioxidant, haemolytic, hypercholesterolemic, lubricant, [c]neonaticide, pesticide, propecic, flavor 5-α reductase inhibitor.
3,7,11,15- Tetramethyl-[R,-[R*,R*-(E)]] Hexadecen-I-ol	Acyclic diterpene alcohol	Antimicrobial, anticancer, anti-inflammatory, hypercholesterolemic, [a]nematicide, pesticide, lubricant, antiandrogenic
9,12-Octadecanoic acid (z, z}-methyl ester	Fatty acid ester	Anti-inflammatory, hypercholesterolemic, cancer-preventive, hepatoprotective, nematicide, insectifuge, antihistamine, antieczemic, anti-acne, 5-alpha reductase inhibitor, antiandrogenic, anti-arthritic, anti-coronary.
l-(+)-Ascorbic acid-2,6- dihexadecanoate gamma-aminobutyric lactam	Vitamin C derivative pyrrolidone	Antioxidant, anti-inflammatory, anti-norciceptive, antibacterial, sperm quality enhancer[c], anticonvulsant, [c]solvent and excipient in veterinary pharmaceutical products
6,10,14-trimethyl-2-pentadecanone	Ketone	Arthropod repellent
Hexadecanoic acid methyl ester	Fatty acid ester	Antioxidant, hypercholesterolemic, lubricant, nematicide, pesticide, hemolytic S-Alpha reductase inhibitor, flavor, antiandrogenic

Sources: Dr. Duke's phytochemical and ethnobotanical databases [online database].
[a] could also be referred to as pharmacological activities.
[b] Suppression of prostaglandin biosynthesis by inhibition of arachidonic acid release in macrophages
[c] ability to hinder conception within the first four hours of life.

TABLE 5.21
Phytochemical Constituents of *Cnidoscolus Aconitifolius* **[21]**

Class of Compounds	Aqueous Extract	Methanol Extract
Alkaloid	+	+
Anthraquinone	-	-
Cardiac glycoside	+	+
Saponins	+	+
Flavonoids	+	+
Carbohydrate	+	+
Tannins	+	+
Phlobatannins	+	-
Steroids	-	-

+, Positive; -, Negative.

in Table 5.19. Anti-inflammatory, antitumor, anti-leucotriene-D4, and anti-androgenic activities were some biological activities recorded for different extracts of *Cnidoscolus aconitifolius*, see Table 5.20. Cardiac glycoside, flavonoids, tannins, and phlobatannins were of moderate occurrence in leaf extracts of the above plant, see Table 5.21.

5.9   CONCLUSION

The presence of phytochemical principles such as tannins, flavonoids, reducing sugars, and carbohydrates is revealed through phytochemical screening. Plants have medicinal values based on their biological and active constituents, and thus, there is an enormous potential for supply of the raw materials for drug synthesis and development. For example, fast screening and evaluating UV peaks of 670 and 410 nm from UV-Vis fingerprints usually correlate with compounds such as chlorophyll and chalcones or aurones, respectively, while a shoulder between 260 and 268 nm depicts a simple phenol. Peaks at 420 nm could represent a carotenoid signature. Furthermore, the hyphenated analytical and spectroscopic techniques offer a valuable research resource to evaluate specific phytochemical structures.

REFERENCES

1. Jiofack, T., Ayissi, C. F., Guedje, N and Kemeuze, V., 2009. Ethnobotany and phytomedicine of the upper Nyong valley forest in Cameroon. *African Journal of Pharmacy and Pharmacology* 3(4): 144–150.
2. Messaoudi, M.I., Filali, H., Tazi, A, and Hakkou, F., 2015. Ethnobotanical survey of healing medicinal plants traditionally used in the main Moroccan cities. *Journal of Pharmacognosy and Phytotherapy* 7(8): 164–182.
3. Kalpesh, N.P., Jayvadan, K., Manish, P.P., Ganesh, C.R. and Hitesh, A.P., 2010. Introduction to hyphenated techniques and their applications in pharmacy. *Pharmaceutical Methods* 1(1): 2–13.

4. Ramesh, B.K., 2017. Application of HPLC and ESI-MS techniques in the analysis of phenolic acids and flavonoids from green leafy vegetables (GLVs). *Journal of Pharmaceutical Analysis* 7(6): 349–364.

5. Anthony, B., Deborah, J and Michael, H., 2012. Value chains of herbal medicines Research needs and key challenges in the context of ethnopharmacology. *Journal of Ethnopharmacology* 140(3): 623–633.

6. Douati, T.E., Kouame, K., Ahou, C., Nyamien, Y., Konan, C., Coulibaly, A., Sidibe, D., Konan, N. Ysidor and Biego, G.H. Marius., 2017. Antioxidants contents of terminalia catappa (Combretaceae) almonds grown in Côte d'Ivoire. *Archives of Current Research International* 10(3): 1–12.

7. Madubuike, U.A. and Rosemary, C.O., 2017. Antimicrobial activity of Nigerian medicinal plants. *Journal of Intercultural Ethnopharmacology* 6(2): 240–259.

8. Joachim, E. and Martin, V., 2000. Applications of hyphenated LC-MS techniques in pharmaceutical analysis. *Biomedical Chromatography* 14(6): 373–383.

9. Adam, P.S. and Peter, W.C., 2006. Isocratic and gradient elution chromatography: A comparison in terms of speed, retention reproducibility and quantitation. *Journal of Chromatography A* 1109: 253–266.

10. Nikolin, B., Imamović, B., Medanhodzić-Vuk, S. and Sober, M., 2004. High perfomance liquid chromatography in pharmaceutical analyses. *Bosnian Journal of Basic Medical Sciences* 4(2):5–9.

11. Deepa, S.C., Sripriya, N. Shankarb and Bangaru, C., 2014. Studies on the Phytochemistry, Spectroscopic characterization and Antibacterial efficacy of *Salicornia brachiata*. *International Journal of Pharmacy and Pharmaceutical Sciences* 6(6): 430–432.

12. Lynn, J.G., 2007. A revision of Paleotropical Plukenetia (Euphorbiaceae) including two New Species from Madagascar. *Systematic Botany* 32(4):780–802.

13. Ndiukwu, M.C and Ejirika, C., 2016. Physical properties of the African walnut (*Tetracarpidium conophorum*) from Nigeria. *Journal of Cogent Food & Agriculture* 2(1): 342–348.

14. Ezealisiji, K.M and Agbo, M.O., 2016. Phytochemical analysis of n-hexane nut extract of *Tetracarpidium conophorum* (Euphorbiaceae) using ultraviolet-visible, fourier transform infrared and gas chromatography mass spectrometry techniques. *Journal of Pharmacognosy and Phytochemistry* 5(6): 332–336.

15. Dr. Duke's Phytochemical and Ethnobotanical Database, 2014. http//www.ars–gov/ duke/ Assessed Thursday 18th Jan. 2015. 20:40 pm.

16. CABI, 2018. *Invasive Species Compendium*. Wallingford, UK: CAB International. http://www.cabi.org/isc./ Assessed Sunday 28th April, 2018, 14:50 pm.

17. Stephen, O.A. and John, A.O.O., 2009. Protective effects of *Annona Muricata* Linn. (Annonaceae) leaf aqueous extract on serum lipid profiles and oxidative stress in hepatocytes of streptozotocin-treated Ddabetic rats. *African Journal of Traditional, Complementary and Alternative Medicines* 6(1): 30–41.

18. Kingsley, C.A., Paulinus, N.O., 2017. Proximate composition, phytochemical analysis, and in vitro antioxidant potentials of extracts of *Annona muricata* (Soursop). *Food Science & Nutrition* 5: 1029–1036.

19. Lex, A.J.T and Barry, E., 2016. *Terminalia catappa* (tropical almond) *Combretaceae* (combretum family). Species Profiles for Pacific Island Agroforestry (www.traditionaltree.org) ver 2.2.

20. Chioma, N.C., Kenneth, M.E and Rukevwe, E.A., 2018. Phyto-chemical characterization of the leaf extracts of *Terminalia catappa* L. (Combretaceae) using ultravioletvisible, fourier transform infrared and gas chromatography-mass- spectroscopic techniques. *Journal of Pharmacognosy and Phytochemistry* 7(1): 2017–2023.

21. Omotoso, A.E., Eseyin, O.O. and Suleiman, M., 2014. Phytochemical analysis of *Cnidoscolus aconitifolius* (Euphorbiaceae) leaf with Spectrometric Techniques. *Nigerian Journal of Pharmaceutical and Applied Science Research* 3(1): 38–49.
22. Atuahene, C.C., Poku-Perempeh, B. and Twun, G., 1999. The nutritive values of chaya leaf meal (*Cnidoscolus aconitifolius*) studies with broilers chickens. *Animal Feed Science and Technology* 77(1999): 163–172.

6 Medicinal Plants from Cameroon for Obstetric and Gynecological Uses

Bazil Ekuh Ewane
Benjamin C. Ozumba
Patience O. Osadebe
Philip F. Uzor
University of Nigeria

Xavier Siwe-Noundou
Rhodes University

CONTENTS

6.1 INTRODUCTION

In many parts of Africa, and in particular Cameroon, there is strong use of traditional remedies. Cameroon has the geographical coordinates of 6.00 N and 12.00 E, located in the Central African region and shares boundary to the West with the Federal Republic of Nigeria, to the North with the Chad Republic, East with the Central African Republic and to the South with Equatorial Guinea, Republic of Congo and Gabon.

Traditional medicine has been used in Cameroon for the treatment of many ailments, and this has gained attention of more than 90% of the indigenous population to make advantage of this natural resource. The WHO estimates that more than 80% of the population in developing countries makes extensive use of traditional medicine especially indigenous people who cannot afford Western pharmaceuticals [1]. Natural and synthetic drug discovery and production approaches have directly been linked to the ethnopharmacological investigations, and over 50% of drugs for clinical use are derivatives of these natural products [2,3]. More effort is required for the documentation of medicinal plant species and their ethnobotanical knowledge. Furthermore, many of these plant species have almost been extinct due to

DOI: 10.1201/9780429506734-6

deforestation and environmental degradation, and there is therefore the need for environmental conservation. In Cameroon, there are more than 250 ethnic groups with a diverse flora heritage which they use for the treatment of many ailments including gyneco-obstetric problems, the most prominent being infertility [4–15].

Moreover, there is also the absence or inaccessibility of modern healthcare services, affordability, cultural/traditional acceptance and, under certain circumstances, effectiveness than their modern counterparts which have caused a large percentage of the population to rely mostly on bio-available plant-based traditional medicines for their primary health care needs including reproductive health [16]. Other shortcomings of drugs include their unavailability, toxicity, efficiency and complicated side effects when compared to natural and bio-available medicinal plants.

In this review work, we present a list of selected plants which are used for the treatment of gynecological and obstetrics problems in Cameroon. Their isolated compounds are also presented.

6.2 ETHNOBOTANICAL USES

More than 80% of the indigenous population, including herbalists and traditional medicinal practitioners, make use of medicinal plant resources for the treatment of gyneco-obstetric problems. As seen in Table 6.1, some of these gyneco-obstetric problems include primary and secondary infertility in both males and females, low sperm count, loss of libido, irregular menstrual cycle and menstrual unrest, threatened abortions, sexually transmitted diseases, difficult delivery, gonorrhea, syphilis, chlamydia, genital itching, sexual weakness, dilation, emmenagogues, amenorrhea, abortifacient, menopausal disorders, menstrual cramps and venereal diseases. Infertility is a prominent ailment that requires more attention especially in the African continent, although it is always wrongly attributed to just women and not men.

Table 6.1 provides the plant families, plant species, vernacular names, major phytochemical compound classes and ethnobotanical applications. A total of 33 plant families were examined containing 73 plant species. These medicinal plant species are used as leaves, fruits, roots, barks or whole plant to prepare recipes for the treatment. They are also taken singly or as a mixture. It is important to note that there are many unexploited traditional medicinal plants that need to be utilized by the indigenous people and pharmaceutical industries. The plant species could be available in gardens, forests, farms or bushes. Plant parts including stems, fruits, seeds, flowers, barks, roots or whole plant could be used for the treatment of various ailments. These medicinal plant parts could be prepared fresh, dried, boiled in water or palm wine, chewed raw or consumed as a mixture with honey, palm oil or kernel oil. At the end of the preparation, the final product is either consumed as a drink, applied to the affected parts as a semi-solid paste or used during bathing. One unknown outcome (whether long-term or short-term effects) is that these end products are not pharmaceutically standardized.

From the analysis of this survey, the medicinal plants belonging to the Asteraceae, Mimosaceae and Moraceae families were the most recorded. The ranking of the Asteraceae family is also in accord with the findings of other investigators [22,29] who undertook their surveys in closer geographic and ecological areas but in different

TABLE 6.1
Cameroonian Plant Species for Traditional Obstetric and Gynecological Uses

S/N	Families	Plant Species	Vernacular Names	Major Phytochemical Compound Class	Ethnobotanical Use + References
1	Acanthaceae	*Eremomastax speciosa* (Hochst).	pinkuidjum(ghomala'a)	Flavonoids, alkaloids, saponins, and tannins [17]	Infertility, urinary infections [18] Infertility, pains of childbirth, menstrual Unrest [17,19].
2		*Justicia insularis* T. Anders	kwe mchie(ghomala'a)	Flavonoids, alkaloids, and glycosides [20]	Infertility, pains of childbirth, menstruation unrest [18,21]
3		*Acanthus montanus* (Nees) Anders	Nsumelab (bechati)	-	Irregular menstruation [18,19,22].
4		*Declipta obanensis* S. Moore	vee gne(ghomala'a)	-	Female infertility [21]
5		*Eremomastax speciosa* (Hochst.) Cufod.	Mbanfen (bechati) Nkwenakam(ghomala'a)	Flavonoids, alkaloids, saponins, and tannins [17]	Regularizes menstrual cycle [22–24]; Infertility, urinary infections [18]; Irregular menstruation, infertility in women [26]; irregular menstruation [23]; generalized pains [9]
6		*Brillantaisia lancifolia* L.	feng gnet(ghomala'a)	-	Female infertility [21]
7		*Dyschoriste perrottetii* (Nees.) O. ktze.	Ntoukesot(bechati)	-	Primary and secondary Infertility [22]
9		*Nelsonia canescens* (Lam.) Spreng.	Ngwanjeu (Bechati)	-	Difficult delivery, threatened abortions [22]
10	Amaryllidaceae	*Crinum distichum* Herb	Lin pedui (ghomala'a)	-	Infertility, amenorrhea; Horns gulps, painful menstruations, sexually transmitted infections [21]
11	Apiaceae	*Centella asiatica* L.	Lin wou wou (ghomala'a)	-	Female infertility [21]

(Continued)

TABLE 6.1 (*Continued*)
Cameroonian Plant Species for Traditional Obstetric and Gynecological Uses

S/N	Families	Plant Species	Vernacular Names	Major Phytochemical Compound Class	Ethnobotanical Use + References
12		*Eryngium foetidum* L.	Lin tie tie (ghomala'a)	-	Abortifacient and emmenagogues [20,27]
13	Apocynaceae	*Rauwolfia vomitoria* Afzel	Nto-aniene (bechati)	Alkaloids	Venereal diseases [22]
14	Araliaceae	*Polyscias fulva* (Hiern) Harms	Pangwi (ghomala'a)	-	Venereal infections [21]
15	Asteraceae	*Ageratum conyzoides* L.	Mré guefah (bechati)	Flavonoids, alkaloids, benzofurans, and terpenes [28]	Infertility [22,23]; Infertility, microbial infections, infections of the genital device [29–34]; abortifacient [27]; quick delivery [9]
16		*Aspilia africana* Linn.	Awagu (bechati)	-	Protracted menstruation [22]
17		*Emilia coccinea* (Sims.) G. Don	Takuteu (ghomela) Mré lapin (biachati)	-	Dysmenorrheal [20,22]
18		*Vernonia guineensis* Benth.	Ayuabem (bechati)	-	Urinary infections Male sterility [22]
19		*Aspilia Africana* (Pers.) C.D. Adams	Sia msou (ghomala'a)	-	Infertility [21]
20		*Bidens pilosa* L.	Kin gne (ghomala'a)	Acetylenic compounds, flavonoids, glycosides, Chalcones, terpenes and essential oils [5,14]	Labor induction [21,37]
21		*Senecio biafrae* (Oliv. & Hiern) J. M.	Doua (ghomala'a)	Dihydroisocoumarins, terpenoids, sesquiterpens, amino acids and mineral salts [37,38]	Painful menstruations [21]

(Continued)

TABLE 6.1 (Continued)
Cameroonian Plant Species for Traditional Obstetric and Gynecological Uses

S/N	Families	Plant Species	Vernacular Names	Major Phytochemical Compound Class	Ethnobotanical Use + References
22		Senecio mannii Hook.	Makoh (ghomala'a)	4-methoxy-6-(11-hydroxystyryl)-pyrone and 4-methoxy-6-(12-hydroxystyryl)-pyrone [40]	gonorrhea, Chlamydia [21]
23		Erigeron floribundus (Kunth) H.B.	Mré gam (ghomala'a)	Flavonoids, alkaloids, saponins, phenols, and tannins [40]	Genital infections [21,45]
24		Spilanthes filicaulis (Schum. &Thonn) C.D. Adams	Pin twe (ghomala'a)	-	Genital infections [17,21]
25		Vernonia amygdalina Del. Cent.	Bekantsu (bechati)		Abortifacient [27,37,42]; menstrual cramps [26]; venereal diseases [43]
26		Terminalia superba		Stigmasterol-3-O-D-glucoside and 4-acetoxy-3,7-dihydroxy-3,6-dimethylhydronaphthalenone [39]	Female infertility [45]
27		Harungana madagascariensis		Xanthones and quinones [45]	Hemorrhoids [40,45]
28		Erythrina sigmoidea		Flavonoids [45]	Syphilis and female sterility [45]; Staphylococcus aureus [46,47]; emmenagogue [48]
29	Basellaceae	Basella alba Linn.	Ntou (bechati)		Infertility difficult [22]; Abortifacient [27,49]
30	Bignoniaceae	Kigelia africana (Lam.) Benth.	Ngong (bechati)	Quinones [45]	Genital itches, Impotence [45]

(Continued)

TABLE 6.1 (*Continued*)
Cameroonian Plant Species for Traditional Obstetric and Gynecological Uses

S/N	Families	Plant Species	Vernacular Names	Major Phytochemical Compound Class	Ethnobotanical Use + References
31		*Spathodea campanulata P.* Beauv.	Nkiete (bechati)	Quinones [45]	Venereal diseases [45]
32	Combretaceae	Pteleopsis hylodendron		3,4-methylenedioxy-3'-O-methyl-4'-O-glycoside ellagic acid, 3,4-methylenedioxy-3'-O-methylellagic Acid, ptelloellagic acid derivative [41]	Treatment of STDs, female sterility [45]; Active against Staphylococcus aureus, S. pyogenes, and Bacillus cereus [50]
33	Commelinaceae	*Commelina benghalensis L.*	Wou wou (bechati)	-	Induce uterine contraction, reduce labor pain, increases dilation, ease childbirth [9]; delivery difficulty [22]
34	Cyatheaceae	*Cyathea maiana*	Ntseu (bechati)		Low sperm count [22]
35		*Pterorhachis zenkeri* Harms.	Ayilalou (bechati)	Terpenoids [44]	Impotence, loss of libido, low sperm count and sexual weakness in men [22]
36	Ericaceae	*Agauria salicifolia* (Comm.) Hook.f. ex Oliv.	Achane (bechati)	-	Sexually transmitted diseases [22]
37	Euphorbiaceae	*Elaephorbia grandifolia* (Haw.) Croizat	Mamcreh (ghomala'a)	-	[21]
38		*Euphorbia tirucalli L.*	Lin kesuh (ghomala'a)	-	Hemorrhoids, gonorrhea; labor induction [1,21];
39		*Phyllanthus amarus* Schum. et Thonn.		-	Urination of blood [22]

(Continued)

TABLE 6.1 (*Continued*)
Cameroonian Plant Species for Traditional Obstetric and Gynecological Uses

S/N	Families	Plant Species	Vernacular Names	Major Phytochemical Compound Class	Ethnobotanical Use + References
40		*Jatropha curcas* L.	Cottonier (ghomala'a)	-	Infertility [20,31,32,52,53]
41	Hypericaceae, Clusiceae or Guttiferae	*Hypericum lanceolatum*		Xanthones and quinines [45]	Venereal diseases and infertility [45]
42	Ixonanthaceae	*Irvingia gabonensis*		Friedelanone, Oleanolic acid, hardwickii acid, 3, 3', 4-tri-O- methylellagic acid, 3,4-di-O- methylellagic acid [54,55]	Gonorrhea [45,54,55]
43	Leguminosae-Papilionoideae	*Millettia Griffoniana*		isoflavonoids and diterpenoids [6]	Sterility, amenorrhea and menopausal disorders [45]
44	Liliacea	*Aloe buettneri* A. Berger	Aloe vera	Glycoside, quinines, coumarins and anthraquinonic derivatives [5]	Infertility, painful menstruations, dysmenorrheal [17,23,55,57]; sterility [57,59,60]
45	Meliaceae	*Pterorhachis zenkeri* Harms.	Ayilalou (bechati)	-	Impotence, loss of libido, low sperm count [45]
46	Mimosaceae	*Albizia lebbeck* (L.) Benth.	Nlame (bechati)	-	Secondary infertility [45]
47		*Allium cepa* L.	Anoussi (ghomala'a)	-	Labor induction [21,37]
48		*Aloe vera* (L.) Burm	Akockdem (bechati)	-	Menstrual cramp [26]
49		*Aloe barbadense* Mill.	Nlamekeu (bechati)	-	Genital itches [22]
50		*Urginea altissima* (Linn.) Bak.	Nlang (bechati)	-	Fibroids, Ovarian cysts, Amnions [22]

(Continued)

TABLE 6.1 (*Continued*)
Cameroonian Plant Species for Traditional Obstetric and Gynecological Uses

S/N	Families	Plant Species	Vernacular Names	Major Phytochemical Compound Class	Ethnobotanical Use + References
51	Moraceae	*Ficus capreaefolia* Del.	Yam (ghomala'a)	-	Syphilis; sexual weakness, syphilis and gonorrhea [21]
52		*Ficus glumosa* (Miq.) Del.	Gah (ghomala'a)	-	Feminine sterility, childbirth [2]
53		*Ficus asperifolia* Miq.	Ntob (bechati)	Flavonoids	Primary and secondary infertility [22]
54		*Ficus sycomorus* L.	Gah douh (ghomala'a)	Galinic tannins, saponins, alkaloids, reducing sugars and aglycon flavons; Flavonoids	Bareness; Sexual weakness, gonorrhea [2]
55	Musaceae	*Musa sapientium* L.	Vuh kedé (ghomala'a)	-	Barrenness, irregular menstrual cycle [21,62]
56	Pentadiplandraceae	*Pentadiplandra brazzeana* Baillon	Allium (ghomala'a)	-	Painful menstruation [21]
57	Periplocaceae	*Mondia whitei* (Hook. f.) Skeel	Nganghelou (bechati)	-	Impotence, low sperm Count and sexual weakness in men [22]
58	Pittosporaceae	*Pittosporum mannii* Hook.f.	Abidong (bechati)	-	Amenorrhea [22]
59	Polygonaceae	*Polygonum nepalense* Meisn.	Aphine (bechati)	-	Foetal malposition, juvenal pregnancy [22]
60		*Rumex crispus* Linn.	Ateuteulukô (bechati)	-	Syphilis [22]
61	Rosaceae	*Prunus africana* (Hook. f.) K. Schum.	Nleh (bechati)	-	Venereal infections [22]
62		*Rubus pinnatus* Willd.	Kakab (bechati)	-	Preparation for child birth [22]
63	Rubiaceae	*Spermacoce princeae* K. Schum	Kom teu (ghomala'a)	-	Syphilis, sexual weakness and gonorrhea [21]; labor induction [37]

(*Continued*)

TABLE 6.1 (Continued)
Cameroonian Plant Species for Traditional Obstetric and Gynecological Uses

S/N	Families	Plant Species	Vernacular Names	Major Phytochemical Compound Class	Ethnobotanical Use + References
64		*Spermacoce saticola K. Schum.*	Tekwentejo (bechati)	-	Juvenal pregnancy [22]
65	Rutaceae	*Zanthoxylum buesgenii Engl.*	Mbem (bechati)	Alkaloids [44]	Loss of libido, low sperm count [26]
66		*Galium asparine Linn.*	Njiekuba (bechati)	-	Gonorrhea [22]
67	Smilacaceae	*Smilax anceps L.*	Khap kape (ghomala'a)	Alkaloids, saponins [2]	Syphilis; female infertility [21]
68	Solanaceae	*Nicotiana tabacum L.*	Depah (ghomala'a)	-	Labor induction [37]; infertility [21]
69		*Solanum aculeastrum Damal.*	Keshimpong (bechati)	-	Amenorrhea Infertility [22]
70		*Solanum torvum Sw.*	Su su dem (ghomala'a)	Flavonoids, alkaloids, saponins, glycosides, and tannins [62,63]	Infertility, microbial infections, genital infections [18,33,64]; Infertility [2]
71	Ternstroemiaceae	*Ternstroemia sp.*	Nkene (bechati)	Saponins, tannins, caffeine and fixed oils.	Primary and secondary infertilities [22]
72	Urticaceae	*Laportea aestuans*[a] (Linn.) Chev.	Mejekephen (bechati)	-	Menopausal disorders [22]
73	Vitaceae	*Cissus quadrangularis L.*	4 cotés (ghomala'a)	-	Gonorrhea and Chlamydia [21]

(-) indicates No report provided.

[a] Bechati and Ghomala'are the commonest traditional language spoken in the regions.

regions. This predominance of plants from the Asteraceae family for the treatment of gyneco-obstetric ailments especially infertility could reflect the worldwide high number of species (19,085) found in this family. These plants are herbaceous species that most often occur as weeds, and their high frequency of use in this study may be related not only to their availability and abundance in this geographical area but also to the similarities of traditional knowledge of people of this area on female infertility treatment with medicinal plants. *Ageratum conyoides L.* and *Bidens pilosa L.* are the most cited species and have the highest frequency of use. This may be due to their effectiveness in action because they contain more bioactive agents that are administered differently and hence different pharmaceutical therapy. The unavoidable fact about these medicinal plants is that most of them are in great danger of extinction maybe due to overconsumption or desertification due to climate change. However, this varies from one community and/or region to another.

6.3 PHARMACOLOGICAL STUDIES AND PHYTOCHEMICAL COMPOSITION

As shown in Table 6.2 and Figure 6.1, these plants contain a variety of bioactive components, and some of them include flavonoids, saponins, tannins, glycosides, xanthones, quinones and quinines. These could be responsible for the bioactivity of these plants. Plants of the same family often produce a similar biochemical compound from their biosynthetic pathway as they possess similar enzymic systems. Majorly,

TABLE 6.2

Isolated Compounds from Some of the Medicinal Plants

Plant Family	Major Phytochemical Component	Genera Reported
Acanthaceae	Flavonoids, alkaloids, saponins, glycoside and tannins	*Eremomastax* and Justicia
Asteraceae	Flavonoids, alkaloids, benzofurans, Acetylenic compounds, glycosides, chalcones, essential oils Terpenes, dihydroisocoumarins, terpenoids, sesquiterpens, amino acids, mineral salts, 4-methoxy-6-(11-hydroxystyryl)--pyrone and 4-methoxy-6-(12-hydroxystyryl) –pyrone, saponins, phenols, tannins, stigmasterol-3-O- -D-glucoside, 4 -acetoxy-3,7-dihydroxy-3,6-dimethylhydronaphthalenone, xanthones and quinones	Ageratum, Bidens, Senecio, Erigeron, Terminalia, Harungana, Erythrina
Bignoniaceae	Quinones	*Kigelia, Spathodea*
Combretaceae	3,4-methylenedioxy-3'-O-methyl-4'-O-glycoside ellagic acid, 3,4-methylenedioxy-3'-O-methylellagic Acid, ptelloellagic acid derivative	Pteleopsis

(Continued)

TABLE 6.2 (*Continued*)
Isolated Compounds from Some of the Medicinal Plants

Plant Family	Major Phytochemical Component	Genera Reported
Cyatheaceae	Terpenoids	*Pterorhachis*
Hypericaceae, Clusiceae or Guttiferae	Xanthones and quinines	Hypericum
Ixonanthaceae	Friedelanone, oleanolic acid, hardwickii acid, 3, 3', 4-tri-O-methylellagic acid, 3,4-di-O- methylellagic acid	Irvingia
Leguminosae-Papilionoideae	Isoflavonoids and diterpenoids	Millettia
Liliacea	Glycoside, quinines, coumarins and anthraquinonic derivatives	Aloe
Moraceae	Flavonoids, Galinic tannins, saponins, alkaloids, reducing sugars and aglycon flavons	*Ficus*
Rutaceae	Alkaloids	*Zanthoxylum*
Smilacaceae	Alkaloids, saponins	Smilax
Solanaceae	Flavonoids, alkaloids, saponins, glycosides and tannins	Solanum
Ternstroemiaceae	Saponins, tannins, caffeine and fixed oils.	*Ternstroemia*

alkaloid and flavonoid classes have highly been reported, and they have been found to be applicable in the treatment of the majority of gyneco-obstetric diseases including infertility and menstrual complications. Some of these plants might have opposite effects in both males and females for the treatment of the same ailment. This could, however, be attributed to the heterogeneous nature of the extract component and their effects on the reproductive hormones.

6.4 CONCLUSION

In conclusion, valuable information has been provided through the study of medicinal plants in Cameroon which is a botanically rich country with diverse medicinal plant species used for the treatment of gyneco-obstetric diseases. Our findings show that these plants are mainly used in crude form. Asteraceae, Mimosaceae and Moraceae were the most recorded family of plants used for the treatment of gynecological and obstetric problems in Cameroon. Fascinating ideas are transferred from older persons who have an in-depth knowledge about the various medicinal plants and their target diseases in their different localities while there is need for the conservation of these plants as some of them are already extinct. This can only be achieved through the education of the population while there is interest to blend traditional and modern medical practices to access the therapeutic effects such as toxicological investigations, hence achieving an extensive productive health.

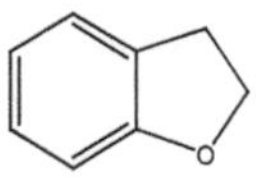

Benzofuran 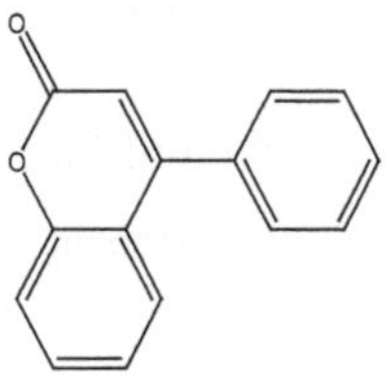   Flavonoid 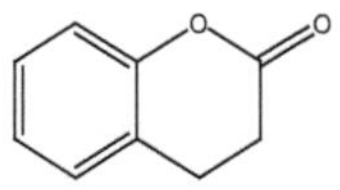   Dihydrocoumarin

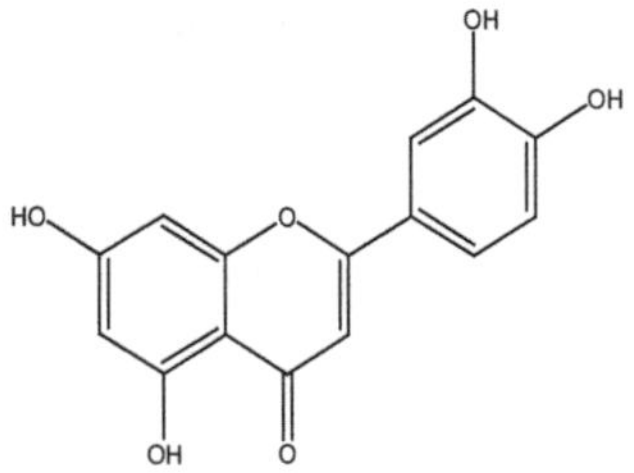

Flavone 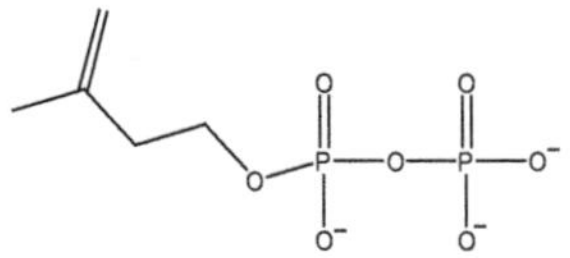   Terpenoid 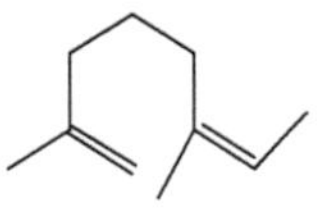   Sesquiterpene

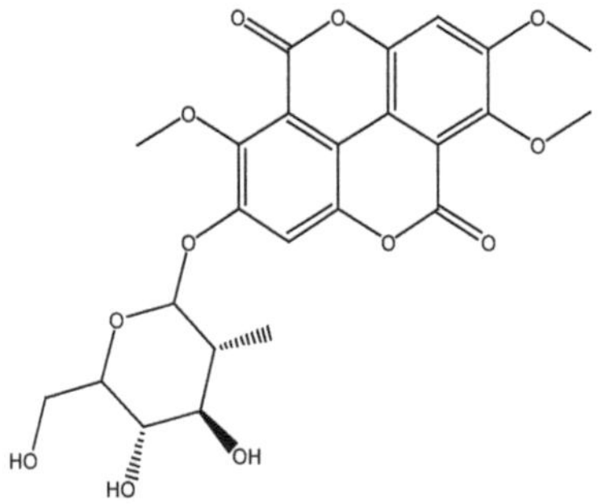

3,4-methylenedioxy-3'-O-methyl-4'-O-glucoside ellagic acid   Aglycon flavone

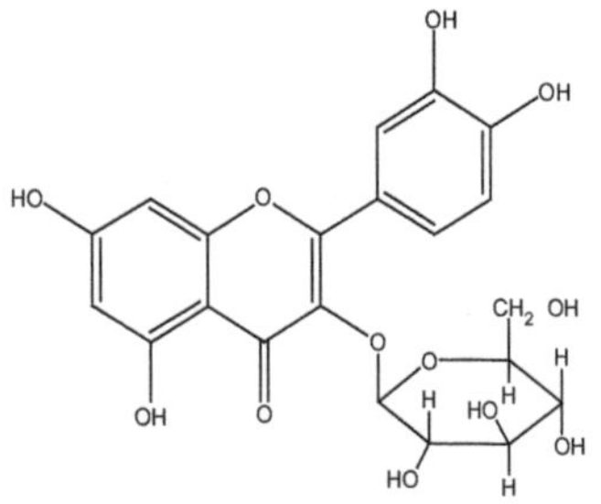

FIGURE 6.1   Structures of some of the isolated compounds.

(Continued)

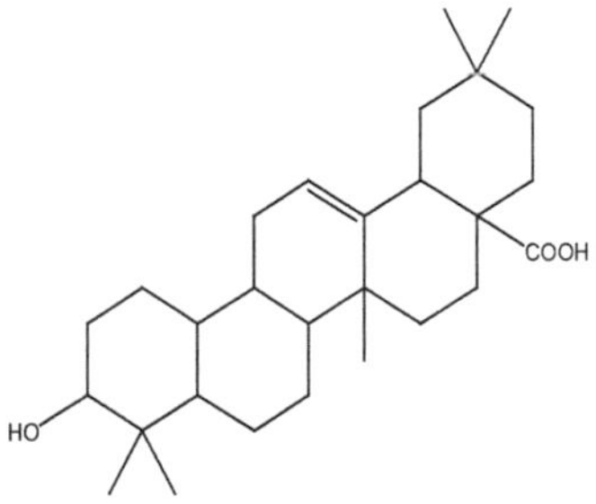

Oleanolic acid

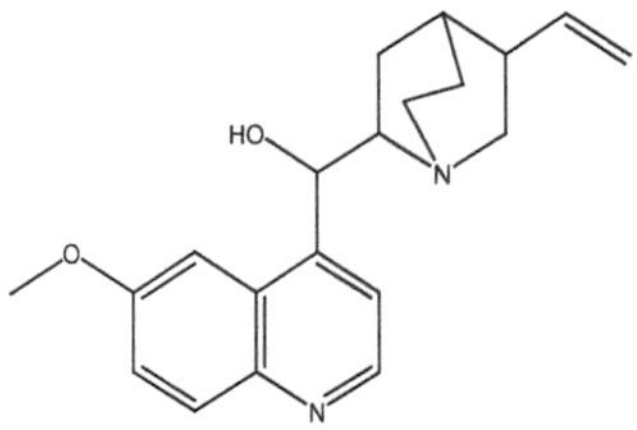

Quinine

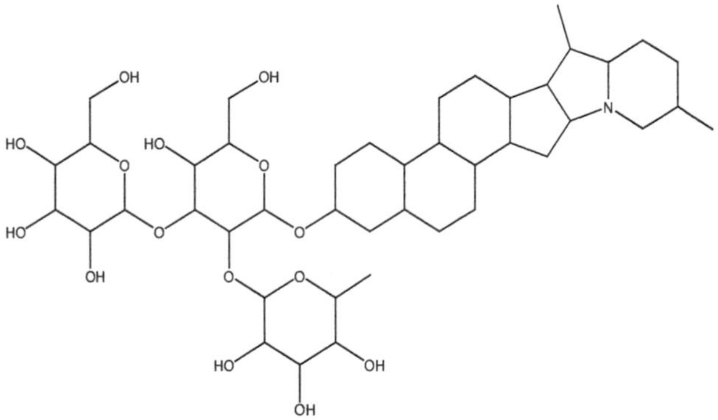

Saponine

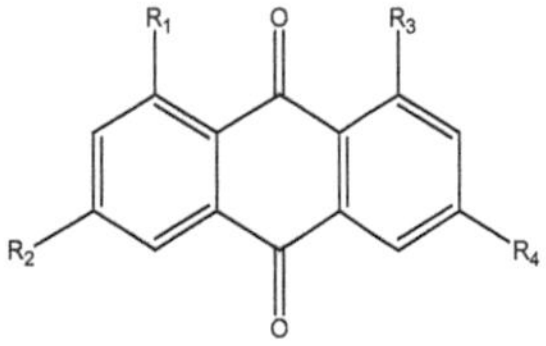

Anthraquinone derivative

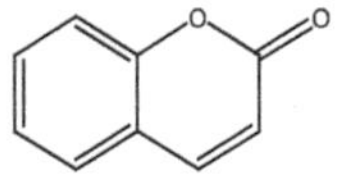

Coumarin 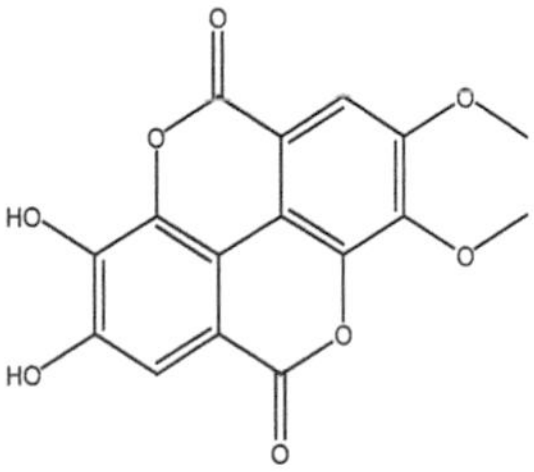   3,4-di-*O*-methylellagic acid 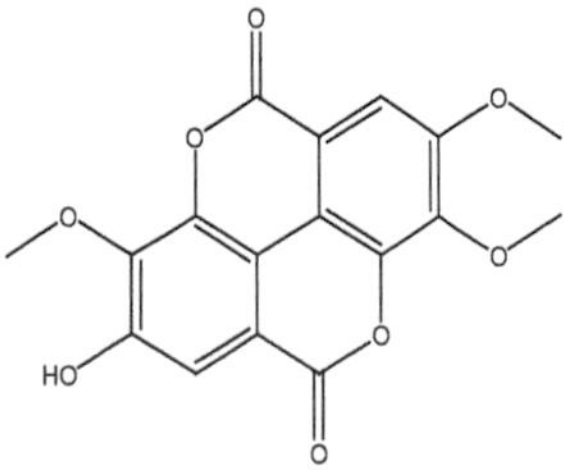   3,3',4-tri-*O*-methylellagic acid

FIGURE 6.1 (Continued)   Structures of some of the isolated compounds.

(Continued)

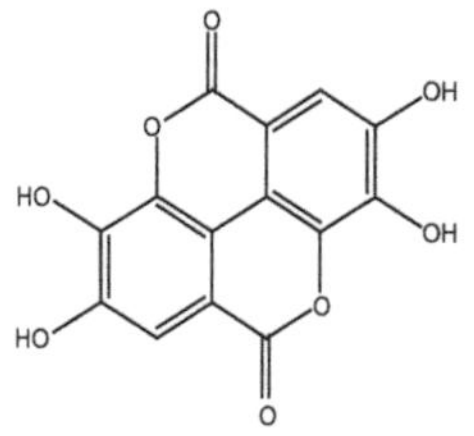

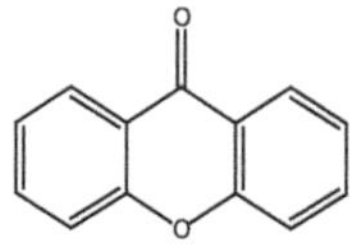

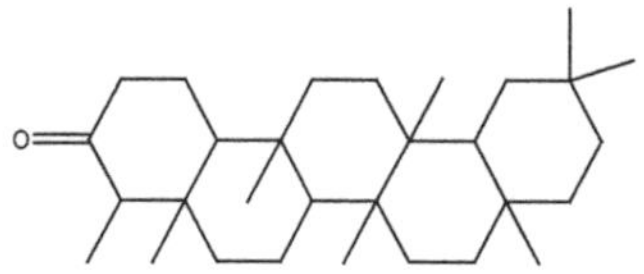

Ellagic acid   Xanthone   Friedalanone

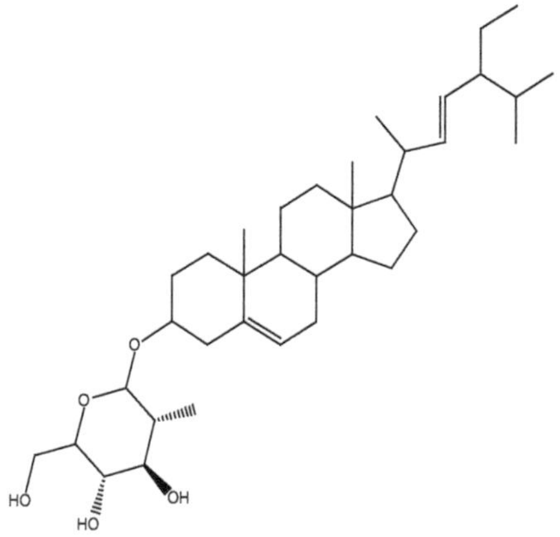

Stigmasterol-3-*O*-D-glucoside 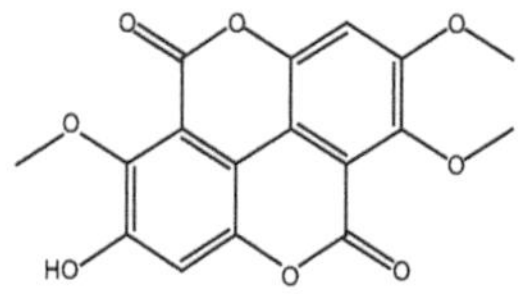   3,4-dimethylenedioxy-3'-*O*-methylellagic acid

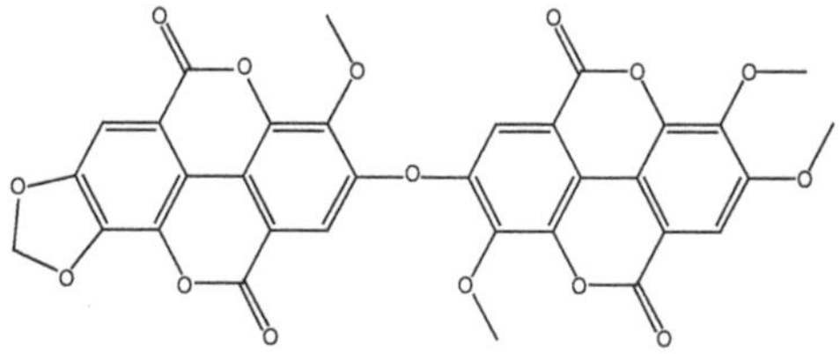

Ptelloellagic acid derivative 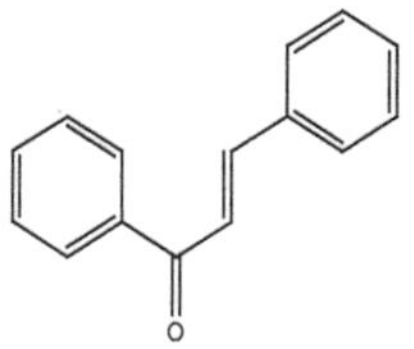   Chalcone

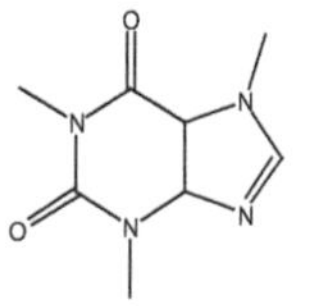

Caffeine

FIGURE 6.1 (*Continued*)   Structures of some of the isolated compounds.

REFERENCES

1. Kamatenesi-Mugisha, M. and Bukenya-Ziraba, R., 2002. Ethnobotanical survey methods to monitor and assess the sustainable harvesting of medicinal plants in Uganda. In: Maunder, M., Clubbe, C., Hankamer, C., Madeleine, G. (Eds.), *Plant Conservation in the Tropics. Perspectives and Practice.* The Royal Botanical Gardens, Kew, Great Britain, pp. 469–482.

2. Fabricant, D.S. and Farnsworth, N.R., 2001. The value of plants used in traditional medicine for drug discovery. *Environmental Health Perspectives (Supplement)* 109: 69–75.

3. Van Wyk, B.E., Van Oudshoorne, B. and Gericke, N., 2002. *Medicinal Plants of South Africa.* Briza Publications, Pretoria, South Africa, p. 336.

4. Guedje, N.M. and Fankap, R., 2001. Utilisations traditionnelles de *Garcinia lucida* et *Garcinia kola* (Clusiaceae) au Cameroun. *Systematics and Geography of Plants.* 71: 747–758.

5. Guedje, N.M., 2002. La gestion des populations d'arbres comme outil pour une exploitation durable des produits forestiers non ligneux : l'exemple de *Garcinia lucida* (sud-Cameroun). The Tropenbos-Cameroon Programm, Kribi, and Université Libre de Bruxelles, Brussels. Tropenbos-Cameroun Series 5, p. xviii + 223.

6. Guedje, N.M., Lejoly, J., Nkongmeneck, B.A. and Jonkers, W.B.J., 2003. Population dynamics of *Garcinia lucida* (Clusiaceae) in Cameroonian Atlantic forests. *Forest Ecology and Management* 177: 231–241.

7. Guedje, N.M., Mouamfon, M., Bigombé Logo, P., Abéga, S.C. and Lejoly, J., 2008. Impact de la gestion socio-économique et technique des forêts communautaires à l'échelle des économies familiales. Cas de Kompia et Kabilone (Est-Cameroun). In: Roulet PA, Assemaker P (eds) *Governance et Environnement en Afrique Centrale: le modèle participatif en question.* Musée Royal de l'Afrique Centrale, Tervuren Belgique, pp. 139–157.

8. Betti, J.L., 2002. Medicinal plants sold in Yaounde markets, Cameroon. *African Study Monographs* 23(3): 47–64.

9. Jiofack, T., Fokunang, C., Kemeuze, V., Fongnzossie, E., Tsabang, N., Nguedjeu, N., Mapongmetsem, P.M. and Nkongmeneck, B.A., 2010. Ethnobotanical uses of medicinal plants of two ethnoecological regions of Cameroon. *International Journal of Medicine and Medical Sciences* 2: 60–79.

10. Dibong, S.D., Mpondo Mpondo, E., Ngoye, A., Kwin, N.F. and Betti, J.L., 2011a. Ethnobotanique et phytomédecine des plantes médicinales vendues sur les marchés de Douala, Cameroun. *Journal of Applied Biosciences* 37: 2496–2407.

11. Dibong, S. D., Mpondo Mpondo, E., Ngoye, A. and Priso, R.J., 2011b. Modalities of exploitation of medicinal plants in Douala region. *American Journal of Food and Nutrition* 1(2): 67–73.

12. Dibong, S.D., Mpondo Mpondo, E., Ngoye, A. and Kwin, N.F., 2011c. Plantes médicinales utilisées par les populations bassa de la région de Douala au Cameroun. *International Journal of Biological and Chemical Sciences* 5(3): 1105–1117.

13. Dibong, S.D., Mpondo Mpondo, E., Ngoye, A. and Priso, R.J., 2011d. Inventory and biodiversity of species edible wild fruits sold in the markets of Douala, Cameroon. *International Journal of Applied Biology and Pharmaceutical Technology* 2(3): 303–311.

14. Din, N., Dibong, S.D., Mpondo E.M., Priso R.J., Kwin, N.F. and Ngoye, A., 2011. Inventory and Identification of Plants Used in the Treatment of Diabetes in Douala Town (Cameroon). *European Journal of Medicinal Plants* 1(3): 60–73.

15. Priso, R.J., Nnanga, J.F., Etame, J., Din, N. and Amougou, A., 2011. Les produits forestiers non ligneux d'origine végétale : valeur et importance dans quelques marchés de la région du Littoral-Cameroon. *Journal of Applied Biosciences* 40: 2715–2726.

16. Vulto, A.G. and Smet, P.A.G.M., 1988. Drugs used in non-orthodox medicine. Meyler's side effects of drugs. *Elsevier: AmsterdamDukes MMG* 11: 999–1005.

17. Oben, J.E., Assi, S.E., Agbor, G.A. and Musoro, D.F., 2006. Effect of Eremomastax speciosaon experimental diarrhea. *African Journal of Traditional Complementary Medicine* 3: 95–100.
18. Adjanohoun, J.E., Aboubakar, N., Dramane, K., Ebot, M.E., Ekpere, J.A. and Enow-Orock, E.G., 1996. *Traditional medicine and pharmacopoeia contribution to ethnobotanical floristic studies in Cameroon.* CNPMS, Porto-Novo Bénin, p. 22.
19. Telefo, P.B., Moundipa, P.F., Tchana, A.F., Tchouanguep, D.C. and Mbiapo, F.T., 1998. Effects of an aqueous extract of Aloe buettneri, Dicliptera verticillata, Hibiscus macranthus, and Justicia insularis on some biochemical and physiological parameters of reproduction in immature female rat. *Journal of Ethnopharmacology* 63: 193–200.
20. Telefo, P.B., Moundipa, P.F. and Tchouanguep, F.M., 2004. Inductive effects of the leaf mixture extract of Aloe buettneri, Justicia insularis, Dicliptera verticillataand Hibicus marcranthuson in vitro production of oestradiol. *Journal of Ethnopharmacology* 90, 225–230.
21. Telefo, P B., Lienoua, L.L., Yemele, M.D., Lemfack, M.C., Mouokeu, C., Goka, C.S., Tagne, S.R. and Moundipa, F.P., 2011. Ethnopharmacological survey of plants used for the treatment of female infertility in Baham, Cameroon. *Journal of Ethnopharmacology* 136: 178–187.
22. Focho, D.A., Ndam, W.T. and Fonge, B.A., 2009. Medicinal plants of Aguambu – Bamumbu in the Lebialem highlands, southwest province of Cameroon. *African Journal of Pharmacy and Pharmacology* 31: 001–013.
23. Lemfack, M.C., 2007. Enquête ethnopharmacologique des plantes utilisées pour le traitement de l'infertilité féminine dans les localités Fossong-Wentcheng et Foto (Menoua, ouest-Cameroun), et activité biologique de la plante la plus utilisée sur la fonction ovarienne de la ratte immature. Thèse de Master of science en Biochimie. Université de Dschang, p. 82.
24. Lienou, L.L., Telefo, P.B., Bayala, B., Yemele, M.D., Lemfack, M.C., Mouokeu, C., Goka, C.S., Tagne, S.R. and Moundipa, F.P., 2010. Effect of ethanolic extract of Senecio biafrae on puberty onset and fertility in immature female rat. *Cameroon Journal of Experimental Biology* 6: 101–109.
25. Ngoufack, M., 2009. Toxicité aiguë, subaiguë d'Eremomastax speciosa, et évaluation de ses effets sur la reproduction chez la ratte. Thèse de master en biochimie, Université de Dschang, Cameroun. 74p.
26. Fonge, B.A., Egbe, E.A., Fongod, A.G.N., Focho, D.A., Tchetcha, D.J., Nkembi, L. and Tacham W.N., 2012. Ethnobotany survey and uses of plants in the Lewoh-Lebang communities in the Lebialem highlands, South West Region, Cameroon. *Journal of Medicinal Plants Research* 6(5): 855–865
27. Duke, J.A., 2003. *Handbook of medicinal species.* CRC Press, London; Emmanuel, N., Claidette, D., 2007. Abortifacient plants of the Buea region their participation in the sexuality of aldoscent girls, Cameroon. *Indian Journal of Traditional Knowledge* 6: 502–507.
28. Adewole, L.O., 2002. Ageratum conyzoïdes L. *Fitoterapia* 73, 1–16.
29. Adjanohoun, E.J., Ake Assi, L., Ahmed, A., Eyme, J., Guindo, S., Kayonga, A., Keita A. and Lebras, M., 1988. *Médecine traditionnelle et Pharmacopée. Contribution aux études botaniques et floristiques au Comores.* Rapport Agence de Coopération Culturelle et Technique, Paris, p. 243.
30. Bouquet, A., 1969. Féticheurs et médecine traditionnelle du Congo (Brazzaville). *Mémoire ORSTOM* 36: 128.
31. Burkill, H.M., 1985. *The useful plants of West Tropical Africa*, vol. 1, 2nd ed. Royal Botanic Garden, Kew, p. 960.
32. Igoli, J.O., Ogaji, O.G., Tor-Anyiin, T.A. and Igoli, N.P., 2005. Traditional medicine practice amongst the Igede people of Nigeria Part II. *African Journal of Traditional Complementary Medicine* 2: 134–152.

33. Iwu, M.M., 1993. *Handbook of Africa medicinal plants*. CRC Press, Boca Raton, Ann Arbor, FL, p. 435.
34. Noumi, E. and Eloumou, M.E.R., 2011. Syphilis ailment: Prevalence and herbal remedies in Ebolowa subdivision (South region, Cameroon). *International Journal of Pharmaceutical and Biomedical Sciences* 2(1): 20–28.
35. Amvam Zollo, H., Kuiate, J.R., Menut, C., Lamaty, G., Bessiere, J.M., Chalchat, J.C. and Garry, R.P., 1995. Aromatic plants of tropical central Africa XX: the occurrence of 1- phenyl-1,3,5-triyne in the essential oil of Bidens pilosa L. from Cameroun. *Flavour and Fragrance Journal* 10: 97–100.
36. Brandao, M.G., Krettli, A., Soares, L.S., Nery, C.G. and Marinuzzi, H.C., 1997. Antiamalarial activity of extracts and fractions from Bidens pilosa and other Bidens species (Asteraceae) correlated with the presence of acetylene and flavonoid compounds. *Journal of Ethnopharmacology* 57: 131–138.
37. Kamatenesi-Mugisha, M., 2004. Medicinal plants used in reproductive health care in western Uganda: documentation, phytochemical and bioactivity evaluation. Ph.D. Thesis. Department of Botany, Makerere University, Kampala, Uganda.
38. Adebooye, O.C., 2004. Solanecio biafrae (Olive and Heirne). In: Jeffery, C., Grubben, G.J.H., Denton, O.A. (Eds.), *Plant Resources of Tropical Africa 2: Vegetables*. PROTA Foundation, Netherlands/Backhuys Publishers, Leiden, Netherlands/CTA Wageningen, Netherlands, pp. 469–471.
39. Tabopda, T.K., Ngoupayo, J., Liu, J., Ali, M.S., Khan, S.N., Ngadjui, B.T. and Luu, B., 2008. α-Glucosidase inhibitors ellagic acid derivatives with immunoinhibitory properties fromTerminalia superba. *Chemical and Pharmaceutical Bulletin* 56(6): 847–850.
40. Ndom, J.C., Mbafor, J.T., Kakam, Z., Happi, N., Vardamides, J.C., Mbaze Meva'a, L., Mpondo, T.N. and Fomum Z.T., 2007. Styrylpyrone glucosides with antimicrobial activity fromSenecio manniiHook. (Asteraceae). *Boletín Latinoamericano y del Caribe de Plantas Medicinales y Aromáticas* 6(3): 73–80.
41. Asongalem, E.A., Foyet, H.S., Ngogang, J., Folefoc, G.N., Dimo, T. and Kamtchouing, P., 2004. Analgesic and antiinflammatory activities of Erigeron floribundus. *Journal of Ethnopharmacology* 91: 301–308.
42. Afegenui, A., 2007. *Ethnobotanical inventory, diseases and chemical screening of some Solanaceae in Tubah sub-division, North West province, Cameroon*. Thesis of degree of masters of Science in botanic. University of Dschang, Cameroon, pp. 41–43.
43. Pamplona-Roger, G.D., 2000. *Encyclopedia of medicinal plants. Education and health library*, vol. 2. Editorial Safeliz, S.L. Spain.
44. Simbo, D.J., 2010. An ethnobotanical survey of medicinal plants in Babungo, Northwest Region, Cameroon. *Journal of Ethnobiology and Ethnomedicine* 6: 8.
45. Ntie-Kang, F., Likowo Lifongo, L., Meva'a Mbaze, L., Ekwelle, N., Owono Owono, L.C., Megnassan, E., Judson, P.N., Sippl, W. and Efange. S.M.N., 2013. Cameroonian medicinal plants: a bioactivity versus ethnobotanical survey and chemotaxonomic classification. *BMC Complementary and Alternative Medicine* 13:147.
46. Kouam, J., Etoa, F.X., Mabeku, L.B.K. and Fomum, Z.T,. 2007. Sigmoidine L, a new antibacterial flavonoid from *Erythrina sigmoidea* (Fabaceae). *Natural Product Communications* 2(11): 1105–1108.
47. Kouam, J., Siwe Noundou X., Mabeku, L.B.K., Lannang, A.M., Choudhary, M.I. and Fomum, Z.T., 2007. Sigmoiside E: A new antibacterial triterpenoid saponin from Erythrina sigmoidea (Hua). *Bulletin of the Chemical Society of Ethiopia* 21(3): 373–378.
48. Kirtikar, K.R. and Basu, B.D., 1946. *Indian medicinal plants*, 3rd ed. Lalit Mohan Basu, Allahabad.

49. Malhi, B.S. and Trivedi, V.P., 1972. Vegetable anti-fertility drugs of India. *Quarterly Journal of Crude Drug Research* 12: 1922–1928.

50. Rahman, A.U., Ngounou, F.N., Choudhary, M.I., Malik, S., Makhmoor, T., Nur-E-Alam, M., Zareen, S., Lontsi, D., Ayafor, J.F. and Sondengam, B.L., 2001. New antioxidant and antimicrobial ellagic acid derivatives from Pteleopsis hylodendron. *Planta Medica* 67: 335–339.

51. Jain, A., Katewa, S.S., Galav, P. and Nag, A., 2008. Some therapeutic uses of biodiversity among the tribal of Rajasthan. *Indian Journal of Traditional Knowledge* 7: 256–262.

52. Igoli, J.O., Tor-Anyiin, T.A., Usman, S.S., Oluma, H.O.A. and Igoli, N.P., 2002. Folk medicines of the lower Benue valley of Nigeria. In: Singh, V.K., Govil, J.N., Hashmi, S., Singh, G. (Eds.), *Recent progress in medicinal plants, 7 ethnomedicine and pharmacognosy II*. Scientific and Technological Publications, USA, pp. 327–338.

53. Kerharo, J. and Gadam, J., 1973. *La pharmacopée sénégalaise traditionnelle-Plantes médicinales et toxiques*. Vigot Frère, Paris, pp. 220–223.

54. Donfack, J.H., Fotso, G.W., Ngameni, B., Tsofack, F.N, Tchoukoua, A., Ambassa, P., Abia, W., Tchana, A.N., Giardina, S., Buonocore, D., Vita Finzi, P., Vidari, G., Marzatico, F., Ngadjui, B.T. and Moundipa, P.F., 2010. In vitro hepatoprotective and antioxidant activities of the crude extract and isolated compounds from *Irvingia gabonensis*. *Asian Journal of Traditional Medicines* 5(3): 79–88.

55. Kuete, V., Wabo, G.F., Ngameni, B., Mbaveng, A.T., Metuno, R., Etoa, F.X., Ngadjui, B. T., Beng, V.P., Meyer, J.J.M. and Lall, N., 2007. Antimicrobial activity of the methanolic extract, fractions and compounds from the stem bark of *Irvingia gabonensis* (Ixonanthaceae). *Journal of Ethnopharmacology* 114(1): 54–60.

56. Kuete, V., 2010. Potential of Cameroonian plants and derived products against microbial infections: a review. *Planta Medica* 76: 1479–1491.

57. Gupta, K., 1972. Aloes compound (a herbal drug) in functional sterility. In: *Proceedings of the XVI Indian Obstetrics and Gynaecology Congress*, New Delhi.

58. Penelope, O., 1994. *Les plantes médicinale: Encyclopédie pratique*. Sélection du Reader's Digest, Paris, p. 192.

59. Bhaduri, A., Ghose, C.R., Bose, A.N., Mosa, B.K. and Basu, U.P., 1968. Antifertility activity of some medicinal plants. *Indian Journal of Experimental Biology* 6: 252–253.

60. Garg, S.K., Saxena, S.K. and Chaudhuri, R.R., 1970. Antifertility screening of plants Part IV. Effects of indigenous plantson early pregnancy in albino rats. *Indian Journal of Medical Research* 58: 1287–1291.

61. Sandabe, U.K., Onyeyili, P.A. and Chibuzo, G.A., 2006. Phytochemical screening and effect of aqueous extract ofFicus sycomorusL. (Moraceae) stembark on muscular activity in laboratory animals. *Journal of Ethnopharmacology* 104: 283–285.

62. Sharma, P.C., 1981. Folklore antifertility plant drugs of Bihar. *Bulletin of Medico-Ethno-Botanical Research* 2: 298–302.

63. Lu, Y., Luo, J., Xu, D., Huang, X. and Kong, L., 2008. Characterization of spirostanol saponins in *Solanum torvum* by high-performance liquid chromatography/evaporative light scattering detector/electrospray ionization with multistage tandem mass spectrometry. *Magnetic Resonance Chemistry* 47: 808–81.

64. Chah, K.F., Muko, K.N. and Oboegbulem, S.I., 2000. Antimicrobial activity of methanolic extract of *Solanum torvum* fruit. *Fitoterapia* 71: 187–189

7 Ondongdong si (*Acmella caulirhiza* Delile), a Medicinal Plant from Cameroon

Tanyi M. Kima and Eleonora D. Goosen
Rhodes University

CONTENTS

DOI: 10.1201/9780429506734-7

7.1   INTRODUCTION

Acmella caulirhiza Delile (Figure 7.1) is one of the most found *Acmella* species in tropical and subtropical Africa. It has numerous medicinal uses in Cameroon, but this chapter will also include its uses in other African countries and Madagascar.

7.2   TAXONOMIC SITUATION

The taxonomic classification of *Acmella* and *Spilanthes* has been rife with uncertainty. The recognized genus, *Spilanthes* Jacq., up till the early 1900s, consisted of two sections, namely *Salivaria* D.C. (19 species) and *Acmella* Rich. DC (44 species) (Table 7.1) [1]. However, it has been shown based on chromosomal and morphological studies that both sections should exist as separate genera. *Spilanthes* (19 species) and *Acmella* Rich ex pers belong to the Asteraceae family (Compositae: subtribe Heliantheae) [2]. Based on these changes, it is not uncommon to find studies using the previous classification resulting in different scientific names for the same plant species. Therefore, it is essential when dealing with these two linked genera to identify other scientific denominations of the species in question.

Since the redefinition of these genera, several species previously classified under *Spilanthes* were moved to *Acmella* Rich. D.C. The genus *Spilanthes* currently consists of six species, while *Acmella* Rich., formerly the smaller section of the *Spilanthes* Jacq. Genus, now includes at least 30 species [3].

The identification and naming of *Acmella caulirhiza* Delile are likewise shrouded with uncertainty and misappropriation. *Acmella caulirhiza* Delile has been described

FIGURE 7.1   *Acmella caulirhiza* Delile (T M Kima).

TABLE 7.1

Homotopic and Heterotopic Synonyms for *Acmella caulirhiza* (Delile) [4–7]

Type of Synonym	Synonyms	Names Marked * and ** Indicate Confidence Levels out of Three Stars for Compounds Found in [6])
Homotopic	*Spilanthes caulirhiza* (Delile) D.C.	**
	Acmella caulirhiza (Delile) DC.	**
Heterotopic	*Ecliptica filicaulis* Schumach. & Thonn.	**
	Spilanthes abyssinica Sch.Bip. ex A. Rich	**
	Spilanthes africana D.C.	**
	Spilanthes filicaulis (Schumach. & Thonn.) C.D. Adams	**
	Spilanthes acmella var. *acmella*	*
	Spilanthes africana var. *africana*	*
	Spilanthes caulirhiza var. *caulirhiza*	*
	Spilanthes caulirhiza var. *madagascariensis* D.C.	*
	Spilanthes mauritiana (Rich. Ex Pers.) D.C.	
	Spilanthes mauritiana f. *mauritania* (DC.) A.H.Moore	
	Acmella mauritiana	
	Blainvillea acmella (L.) Philipson	
	Blainvillea dall-vedovae A. Terracc.	
	Blainvillea dichotoma (Murray) Stewart	
	Blainvillea gayana auct.	
	Blainvillea latifolia (L. f.) DC.	
	Blainvillea rhomboidea Cass	
	Eclipta latifolia L. f.	
	Spilanthes acmella (L.)	
	Spilanthes acmella (L.) Murray	
	Spilanthes acmella (L.) Murray var. *acmella*	
	Spilanthes mauritiana f. *madagascariensis* (DC.) A.H.Moore	
	Verbesina acmella (L.)	
	Spilanthes acmella Murr.	Unresolved

and studied using various names [4–7]. However, the currently accepted name is now *Acmella caulirhiza* Delile.

7.3 PLANT MORPHOLOGY

Following their previous classification under the same genus, members of both genera share similar characteristics. For example, they both present themselves as prostate to decumbent herbaceous perennials. Some other similarities between the members of both genera are that they both possess conical-shaped receptacles, while

members of both genera lack bracts on peduncles and inflorescence axis. They also
lack both glands of pales and glands of corollas. The relative length of corollas to
pales is approximately but not precisely equal in members of both genera. Both
genera's achenes are wingless, while the pubescence of the achene margins is ciliate
with double hairs [2].

However, similarly, they bear enough differences to be widely accepted as
separate genera. The chromosome number in *Spilanthes* is 16, and in A*cmella,* it
is 12 or 13. Morphologically, in *Spilanthes,* the pappus consists of 1, 2 subequal
or unequal awns. The outer achenes may have three awns, while some members
of *Acmella* possess one to several subequal or unequal bristles and others lack
them altogether. The arrangement of their inflorescence alone cannot differentiate
members of both genera. However, the members of *Spilanthes* only have discoid
heads, while 26 *Acmella* species possess radiate heads, nine have discoid heads,
and four have both [8]. Discoid members in both species often have white to pur-
plish or greenish-white corolla, while the corollas in radiate members of *Acmella*
are orange-yellow to yellow. All members of *Spilanthes,* except *Spilanthes costata,*
lack petioles, broaden toward the apex of the pale, and possess four angled mono-
morphic achenes with two, four, or eight vascular strands. In *Acmella*, the leaves
are petiolate, while the pale narrows toward the apex. The achenes of *Acmella*
plants are dimorphic with three angled outer series and two angled inner series
achenes. The achenes of *Spilanthes* have well-defined cork-like margins, while
they are absent in members of *Acmella* [2].

Acmella caulirhiza Delile is an herbaceous perennial with decumbent to repent
stems. It is one of the radiate species of this genus. It possesses a pale yellow to yellow corolla with inconspicuous ray florets which do not or only barely exceed the
phyllaries. The phyllaries of the outer series in this species are longer than the inner
series. This plant flowers all year and can be found in moist areas around river and
stream banks in tropical and sub-tropical Africa. In the *Acmella* genus, *Acmella caulirhiza* Delile is most closely related to *Acmella calva*. However, most calva subtypes
have discoid heads, and they are mostly found in southern to south-east Asia [8].

7.4   ETHNOMEDICINAL USES OF *ACMELLA CAULIRHIZA* DELILE

Acmella caulirhiza Delile and its synonyms are used extensively in Cameroon and
other tropical and sub-tropical areas on the African continent as ethnomedicines to
treat both humans and livestock. The plant's ethnomedicinal uses in these different
areas are included in the following tables according to the different body systems.

7.4.1   DISEASES OF THE EYES, NOSE, MOUTH, THROAT, AND LUNGS

Acmella caulirhiza Delile is used in the treatment of tonsillitis in Ethiopia and
Rwanda. It is also used to treat toothache in Cameroon, Kenya, Ethiopia, Uganda,
Nigeria, and South Africa. It is not clear whether the treatment plays an analgesic or
antibacterial role or both. Coughs and colds are treated with this plant in Cameroon
and Gabon, and people in Rwanda use it to treat Pneumonia (Tables 7.2 and 7.3).

TABLE 7.2
Diseases of the Eyes, Nose, Mouth, Throat, and Lungs

Disease	People and Place	Plant Name	Alternative Names	Plant Part	Preparation	Administration	Reference
Sore throat and throat infections	Bini people from Benin City, Edo State, Nigeria	*Spilanthes filicaulis* (Schum and Thom) J.C.D. Adams	"Brazil cress," "Ehie edo oto."	Flowers	Mix with either three seeds of alligator pepper or dry gin and chewed.	Swallow juice from chewing twice a day	[9]
Oral thrush	Marakwet Community, Kenya	*Spilanthes mauritiana* (Pers.) D.C.	Not provided	Whole plant	Adults: crushed Children: crushed and mixed with water or milk	Adults and Children: Use as a mouthwash	[10]
Tonsillitis	Jimma, Ethiopia	*Spilanthes mauritiana* (Rich.ex.Pers)DC	Yemder berbere	Flowers collected in odd numbers.	They are chewed to form an extract.	Swallow extract. Continue medication until cured	[11]
Tonsillitis	Rwanda	*Spilanthes mauritiana* (Pers.) D.C.	Gashegenyura	Not provided	Not provided	Not provided	[12]
Tonsillitis	Indigenous People, Gozamin Wereda, East Gojjam Zon, Amhara Region, Ethiopia	*Acmella caulirhiza* Del.	Not provided	Leaf	Extract	Oral: one spoon	[13]
Tonsillitis	Sheko people Ethiopia	*Acmella caulirhiza* Delile	Not provided	Flowers	Not provided	Not provided	[14]

(Continued)

TABLE 7.2 (Continued)

Diseases of the Eyes, Nose, Mouth, Throat, and Lungs

Disease	People and Place	Plant Name	Alternative Names	Plant Part	Preparation	Administration	Reference
Tonsillitis	Gamo people, Bonke Woreda, SNNPR, Ethiopia	Acmella caulirhiza	Fach facho	Fresh flowers	Adult: chewed with water Child: chewed by mother	Adult: Oral Child: Oral form mother's mouth	[15]
Tonsillitis	Chencha District; Gamo Gofa, Ethiopia	Acmella caulirhiza Del					[16]
Tonsillitis	Orombo people, Ghimbi District, Southwest Ethiopia	Acmella caulirhiza Delile	Gutichaa	Flowers	Chew	Spit on tonsils of patient	[17]
Tonsillitis	Selale mountain ridges, North Shoa, Ethiopia	Acmella caulirhiza Delile	Gutichaa	Flowers	Chew	Spit on tonsils of patient	[18]
Tonsillitis	Maale and Ari ethnic communities, southern Ethiopia	Acmella caulirhiza Delile	Yemdir Berbere	Flowers	Not provided	Not provided	[19]
Toothache	Marakwet Community, Kenya	Spilanthes mauritiana (Pers.) D.C.	Not provided	Whole plants.	Crushed.	Placed directly on the affected teeth	[10]
Toothache	Foulbé women from the Kambo and Longmayagui watersheds near Douala, Cameroon	Acmella caulirhiza L.	Yeux de poule, Quité gortogal	Whole plants	Macher	Apply crush on tooth decay	[20]
Toothache	Mount Cameroon	Spilanthes filicaulis	Not provided	Flowers	Not provided	Not provided	[21]
Toothache	Maale and Ari ethnic communities, southern Ethiopia	Acmella caulirhiza Delile	Yemdir Berbere	Flowers	Not provided	Not provided	[19]

(Continued)

TABLE 7.2 (*Continued*)

Diseases of the Eyes, Nose, Mouth, Throat, and Lungs

Disease	People and Place	Plant Name	Alternative Names	Plant Part	Preparation	Administration	Reference
Toothache	Indigenous People, Gozamin Wereda, East Gojjam Zon, Amhara Region, Ethiopia	*Acmella caulirhiza* Delile	Yemidir berbere	Flower is	Chewed	chewed for few minutes and spit out	[13]
Toothache	Takkad People of Kaduna State, Nigeria	*Spilanthes filicaulis* (Schumach.&Thonn.) C.D.Adam	Sangkali	Fresh leaves mixed with leaves of *Tapinanthus dodoneifolius*	A paste is made	A small quantity of paste is placed on the affected tooth for 5 min. Pain ceases, symptoms disappear.	.[22]
Toothache	Nkoung–Khi Division West Region Cameroon	*Spilanthes africana* L.	Not provided	Whole plant	decoction of 100grs	drink a lukewarm glass two times per day, morning and evening	[23]
Toothache	Kenya	*Spilanthes mauritiana*	Not provided	Not provided	Not provided	Not provided	[24].
Toothache	South West Region of Cameroon	*Acmella caulirhiza*	Medmekube, Bakossi	Fruits	Maceration	Not provided	[25].
Toothache	Baka pygmies, Dja biosphere, Cameroon	*Acmella caulirhiza*	Not provided	Leaves	Pound to a paste	Apply directly to the tooth	[26]
Toothache	Amhara people, Mecha Wereda, West Gojjam zone, Amhara region, Ethiopia	*Acmella caulirhiza*	Yemidir berberie	Two flowers	Chew	Keep on the tooth as long as pain is experienced	[27][28]
Toothache	IsiXhosa people, South Africa	*Spilanthes mauritiana* D.C.	Not provided	Flowers	Chew	About a minute until the mouth tingles and becomes numb. Numbness disappears after 20 min, but pain does not reappear.	[29]

(*Continued*)

TABLE 7.2 (*Continued*)
Diseases of the Eyes, Nose, Mouth, Throat, and Lungs

Disease	People and Place	Plant Name	Alternative Names	Plant Part	Preparation	Administration	Reference
Toothache	IsiZulu South Africa	*Spilanthes mauritiana* DC	Isisilili	Leaves	Powdered	Moisten powder and put into tooth cavities to relieve pain	[29]
Toothache	Ngai Subcounty, Apac District, Northern Uganda	*Acmella caulirhiza* Delile	Arere	Flowers, roots, and leaves	Crushed	Placed on tooth	[30]
Tirs himem: Dental pain with or without swelling, possibly due to tooth decay	Berta ethnic group Assosa Zone, Benishangul-Gumuz regional state, mid-west Ethiopia	*Acmella caulirhiza* Del	Eisegne andewu	Root	Ground	Put in-between teeth	[31]
Pneumonia	Rwanda	*Spilanthes mauritiana* (Pers.) D.C.	Gashegenyura	Not provided	Not provided	Not provided	[12]
Colds	Foulbé women from the Kambo and Longmayagui watersheds near Douala, Cameroun	*Acmella caulirhiza* L.	Yeux de poule, Quité gortogal	Whole plants	washed and crushed to extract the juice	One spoonful three times per day	[20]

(Continued)

TABLE 7.2 (*Continued*)
Diseases of the Eyes, Nose, Mouth, Throat, and Lungs

Disease	People and Place	Plant Name	Alternative Names	Plant Part	Preparation	Administration	Reference
Colds	Masango from Ibagha and Muyanama, Ngounie province, Gabon	*Spilanthes caulirhiza* (Delile) D.C.	Kungui	Leaves	Chewed	Swallow the juice	[32]
Cough	Mount Cameroon	*Spilanthes filicaulis*	Not provided	Leaves, flowers	Not provided	Not provided	[21]
Sore mouth	IsiZulu South Africa	*Spilanthes mauritiana* DC	Isisitili	Leaves	Powdered	Children: moisten the powder and rub on the mouth.	[29]
Mouth ulcers	Madagascar	*Acmella caulirhiza* Del	Acmelle	whole plant	Not provided	Not provided	[33]
Halitosis	Cameroon	*Spilanthes africana*	Not provided	Not provided	Not provided	Mouthwash with a peppermint taste	[34]
Eyes of farm animals	Southern Ethiopia	*Acmella caulirhiza* Delile	Hajilo	Flowers	Pounded, mixed, and squeezed with the fresh vegetative parts of *Basella alba* L	The liquid mixture is administered three times per day to the eyes of cattle, sheep, horses, and goats	[35]
Eye infection	Northern sector of Kibale National Park, Uganda	*Acmella caulirhiza* Delile	Ensoimya	Fresh roots	Squeeze	applying the liquid directly to infected eyes	[36]
Eye injury	Forests, Awi Zone, north-western Ethiopia	*Acmella caulirhiza*	Yemidir berberie	Fresh whole plant	crushed and squeezed with water	The liquid is then dropped into the eye	[27]

7.4.2　Diseases of the Central Nervous System

TABLE 7.3
Diseases of the Central Nervous System

Disease	People and Place	Plant Name	Alternative Names	Plant Part	Preparation	Administration	Reference
Neuralgia	Bui Division, northwest region of Cameroon	*Acmella caulirhiza*	Sishursheshiv	Leaves	200 g combined with *Crinum jagus* L. (200 g leaves), *Eremomastax speciose* (200 g leaves), and *Gladiolus gregarius* (200 g leaves and 50 g bulbs) boiled together in 5 L of water for 15 min.	Drink 125 mL of the cooled decoction three times per day	[37]
Hemiplegia	Bui Division, northwest region of Cameroon	*Acmella caulirhiza*	Sishursheshiv	Leaves	150 g, *Desmodium intortum* (200 leaves and roots), *Eremomastax speciose* (200 g leaves), *Lannea kerstingii* (150 g leaves), and *Ricinus communis* (200 g leaves) are boiled together in 6 L of water for 15 min.	Drink 125 mL of the cooled decoction three times per day for 12 months	[37]
Local anesthetic	Mount Cameroon Region, Cameroon	*Spilanthes filicaulis*	Not provided	Leaves and flowers	Not provided	Rub on skin.	[21]
Local anesthetic to prepare for craniotomy	Marakwet Community, Kenya	*Spilanthes mauritiana* (Pers.) D.C.	Not provided	Whole plant	Crushed and soaked in water And water macerate	Applied on the area to be operated before craniotomy and for 2 days after surgery Water macerate administered on the evening before the surgery	[10]

(Continued)

TABLE 7.3 (*Continued*)
Diseases of the Central Nervous System

Disease	People and Place	Plant Name	Alternative Names	Plant Part	Preparation	Administration	Reference
Epilepsy	Fongo-Tongo village, Western Province, Cameroon	*Acmella caulirhiza* Delile	Pan touh	Whole plant	Five whole plants, a handful of young leaves from *Solanecio manii*, half a handful of leaves of *Dissotis rotundifolia*, one handful of leaves of *Eremomastax*, and 500 g of *Bridelia* stem bark mixed, crushed, and macerated in water for 24 h	250 mL three times every morning as an enema for 1 month	[38]
Epilepsy	Fongo-Tongo village, Western Province, Cameroon	*Acmella caulirhiza* Delile	Pan touh	Whole Plant	One handful with *Piper capense* (one handful of leaves), *Eremomastax speciosa* (one handful of leaves), and *Biophytum persianum* (one handful of the whole plant) boiled in 5 L of water	Drink every morning. After 30 months of treatment, 10 years of recovery	[38]
Headache	Mount Cameroon Region, Cameroon	*Spilanthes filicaulis*	Not provided	Not provided	Not provided	Not provided	[21]

7.4.3   INFECTIOUS DISEASES

Reports on the ethnomedicinal use of *Acmella caulirhiza* Delile to treat infectious diseases are limited to a few diseases caused by viruses. Examples are dysentery (Cameroon), herpes (Ethiopia), typhoid fever (Cameroon), malaria (Kenya and Rwanda), various infections in animals (Ethiopia), and Guinea worm (Cameroon) (Table 7.4).

7.4.4   DISEASES OF THE GASTROINTESTINAL TRACT

The ethnomedicinal use of *Acmella caulirhiza* Delile to treat diseases of the gastrointestinal tract (Table 7.5) could be due to infections caused by microorganisms. The most documented treatments are for peptic ulcers (Cameroon) and diarrhea (Cameroon, Kenya, and Nigeria). Nonspecific ailments such as side pain (Cameroon) and bloating in livestock (Ethiopia) are also mentioned in the literature.

7.4.5   DISEASES OF THE HEART AND VASCULAR SYSTEM

There are few reports in the literature of the ethnomedicinal use of *Acmella caulirhiza* Delile in the literature. They include chest pain (Cameroon), angina (Cameroon), use as a blood coagulant (Cameroon), and treatment of the swelling of feet (Madagascar) that could be ascribed to heart failure (Table 7.6).

7.4.6   SKIN DISORDERS

Skin disorders treated with *Acmella caulirhiza* Delile include eczema (Cameroon) and some conditions that could be caused by infections, for example, boils (Cameroon), pus discharge (South Africa), and wound dressing (Uganda) (Table 7.7).

7.4.7   DISEASES OF THE UROGENITAL SYSTEM

Impotence (Uganda), prostatic disease (Cameroon), and genital infections (Cameroon) are also treated with *Acmella caulirhiza* Delile (Table 7.8).

7.4.8   CHILDBIRTH AND CHILD CARE IN HUMANS AND ANIMALS

Acmella caulirhiza Delile is used in Cameroon to facilitate delivery and in Uganda to induce labor. It is used to treat nappy rash in Cameroon, for fontanel closing in Gabon, and as a lactation stimulant in Ethiopia (Table 7.9).

7.4.9   NONCOMMUNICABLE DISEASES

Fibromyoma and cancer, in general, are treated with *Acmella caulirhiza* Delile in Cameroon. It is also used to treat diabetes in Cameroon and Kenya (Table 7.10).

TABLE 7.4
Infectious Diseases

Disease	People and Place	Plant Name	Alternative Names	Plant Part	Preparation	Administration	Reference
Dysentery	Mbalmayo Region, Central Province, Cameroon	*Spilanthes filicaulis* Schum. et Thonn. C. D. Adams	Ododongsi	Whole plant	Powdered and mixed with the powder of *Aframomum letestuanum*	Over 2 weeks	[39]
Shererit (Skin rash similar to Herpes)	Bertha ethnic group, Assosa Zone, Benishangul-Gumuz regional state, mid-west Ethiopia	*Acmella caulirhiza*	Gutecha	Leaves	Ground, mixed with sesame oil	applied to the affected area	[31]
Typhoid fever	the Bamnoutos Division in Western Cameroon	*Acmella caulirhiza* Delile	Ehengui, Pento'o, Pentouoare	Leaves and stems	Concoctions, crushed extracts, or decoctions	Oral	[40]
Typhoid fever	South West Region of Cameroon	*Acmella caulirhiza*	Medmekube, Bakossi	Fruits	Macerate	Not provided	[26,42]
Malaria (control of *Anopheles* mosquito)	Kenya	*Spilanthes mauritiana*	Not provided	Not provided	Not provided	Not provided	[24]
Malaria	Rwanda	*Spilanthes mauritiana* (Pers.) D.C.	Gashegenyura	Not provided	Not provided	Not provided	[12],
Veterinary: Actinobacillosis, Bugunch	Ethiopia	*Acmella caulirhiza* Del	Not provided	Leaves	Roasted, ground, and mixed with salt	They are applied to the affected part.	[41]
Guinea worm	Mount Cameroon Region, Cameroon	*Spilanthes filicaulis*	Not provided	Leaves and flowers	Not provided	Not provided	[21]

TABLE 7.5
Diseases of the Gastrointestinal Tract

Disease	People and Place	Plant Name	Alternative Names	Plant Part	Preparation	Administration	Reference
Diarrhea and bloody stool	Mbalmayo Region, Central Province, Cameroon	*Spilanthes filicaulis* Schum. et Thonn. C. D. Adams,	Ododongsi	Leaves	Macerate one handful in 1 L of water	Drink one glass 4–5 times per day	[40,44]
Liver disease (Dhier Tiru)	Bale Mountains National Park, South Eastern Ethiopia	*Acmella caulirhiza* Del	Not provided	Whole plant fresh or dried	concocted, crushed, powdered, mixed with water and sugar	Oral	[43]
Diarrhea	Kenya	*Spilanthes mauritiana*	Not provided	Not provided	Not provided	Not provided	[24]
Infant Diarrhoea	Nigeria	*Spilanthes filicaulis*	Osete, Oliturare	Not provided	Soup prepared with dried fish	One small spoon three times per day	[44]
Stomach problems	Mount Cameroon Region, Cameroon	*Spilanthes filicaulis*	Not provided	Leaves and flowers	Not provided	Not provided	[21]
Emetic	Bini people, Benin City, Edo State, Nigeria	*Spilanthes filicaulis (Schum and Thom) J.C.D. Adams.*	Brazil cress, Ehie edo oto	Leaves	Mixed with alum	Not provided	[9]
Peptic ulcers	Cameroon	*Spilanthes filicaulis (Schum et Thon) C.D. Adams (Asteraceae),*	Not provided	Whole plant	Infusion	Not provided	[45]

(Continued)

TABLE 7.5 (*Continued*)
Diseases of the Gastrointestinal Tract

Disease	People and Place	Plant Name	Alternative Names	Plant Part	Preparation	Administration	Reference
Peptic ulcer	Cameroon	*Spilanthes filicaulis* Jacq.	Laq gap (Bangangte), Ondongdong si (Yaounde), Pang touoh (Dschang);	Leaves or whole plant	A paste of 200 g of plant material with five fruits of *Capsicum frutescens* kneaded in 1 L of water and filtered	One full glass twice a day	[46]
Enema for side pain	Mount Cameroon Region, Cameroon	*Spilanthes filicaulis*	Not provided	Leaves and flowers	Not provided	Not provided	[21]
Bloating in livestock	Ethiopia	*Acmella caulirhiza* Delile	Hajilo	Flowers	pounded, mixed, and squeezed with the fresh vegetative parts of *Basella alba* L	The liquid mixture is administered orally to treat sheep, goats, cattle, and horses that are bloated. Treatment is repeated until recovery	[35]

TABLE 7.6
Diseases of the Heart and Vascular System

Disease	People and Place	Plant Name	Alternative names	Plant Part	Preparation	Administration	Reference
Blood coagulant	Mount Cameroon Region, Cameroon.	*Spilanthes filicaulis*	Not provided	Leaves and flowers	Not provided	Not provided	[21]
Chest pain	Mount Cameroon Region, Cameroon.	*Spilanthes filicaulis*	Not provided	Leaves and flowers	Not provided	Not provided	[21]
Headaches and angina	Santchou, Cameroon	*Spilanthes filicaulis* C.D. Adams	Not provided	Leaves and stems	Not provided	Not provided	[47]
Swollen feet	Madagascar	*Acmella caulirhiza* Del.	Acmelle	Leaves	Not provided	Not provided	[33,48]

TABLE 7.7
Skin Disorders

Disease	People and Place	Plant Name	Alternative Names	Plant Part	Preparation	Administration	Reference
Boils	South West Region of Cameroon	*Acmella caulirhiza*	Medmekube, Bakossi	Fruit	Macerate	Place on boils	[25]
Wound dressing	Ngai Subcounty, Apac District, Northern Uganda	*Acmella caulirhiza* Delile	Arere	Flowers, roots, and leaves	Crushed	Placed on wound	[30]
Pyorrhea (pus discharge)	IsiXhosa people, South Africa	*Spilanthes mauritiana* D.C.	Not provided	Flowers	Chew	Not provided	[29]
Eczema	Mount Cameroon Region	*Spilanthes filicaulis*	Not provided	Leaves and flowers	Not provided	Not provided	[21]
Eczema	South-West Region of Cameroon	*Spilanthes filicaulis*		Whole plant	Decoction	Not provided	[25]

TABLE 7.8
Diseases of the Urogenital System

Disease	People and Place	Plant Name	Alternative Names	Plant Part	Preparation	Administration	Reference
Genital infections	Baham, Cameroon	*Spilanthes filicaulis* (Schumach.&Thonn.) C.D.Adam	Pin twe	Whole Plant	Macerate	Not provided	[49]
Prostatic disease	Cameroon	*Acmella caulirhiza* Delile	Tso guip	Dried leaves: two handfuls flowers	Not provided	Not provided	[50]
Impotence	Uganda	*Spilanthes filicaulis*	Not provided		Decoction mixed with the stem bark and trunk of *Nuxia floribunda*	Not provided	[51]

TABLE 7.9

Childbirth and Childcare in Humans and Animals

Disease	People and Place	Plant Name	Alternative Names	Plant Part	Preparation	Administration	Reference
Lactation stimulant in livestock.	Jimma zone, Ethiopia	*Acmella caulirhiza* Delile:	Yemdir beriberi	Not provided	Powder separately and then add to hot water to make a decoction	Oral	[52]
Fontanel closing (fesse rouge)	Rural areas of Gabon	*Acmella caulirhiza* Delile	Andong si	leaves	Not provided	Not provided	[53]
Nappy rash (red buttocks)	Foulbé women from the Kambo and Longmayagui watersheds near Douala, Cameroun	*Acmella caulirhiza* L.	Yeux de poule, Quité gortogal	Dried leaves	Crush, sift and mix with other ingredients (not provided)	Oral: One tablespoon after the child has been bathed.	[20]
Facilitate Delivery	Menoua division, West Cameroon	*Spilanthes filicaulis* (S. &T.) Adams.	Pantoueh	Whole Plant	Not provided	Not provided	[54]
Labor induction	Runyaruguru people, Western Uganda	*Acmella caulirhiza* Del. (Syn. *Spilanthes africana* D.C.	Esiimwe	Leaves and flowers	Squeeze by hand	Oral	[55]

TABLE 7.10
Noncommunicable Diseases

Disease	People and Place	Plant Name	Alternative Names	Plant Part	Preparation	Administration	Reference
Cancer	South-West Region of Cameroon	*Spilanthes filicaulis*	Not provided	Whole plant	Decoction	Not provided	[25]
Fibromyoma	South Cameroon	*Acmella caulirhiza* Delile	Andong si	*Acmella caulirhiza* Delile: 2 leafy plants *Clerodendron volubile* (Beyem elok) 1 handful of young plants, *Momordica foetida*: 1 handful of leaves, *Solanum lycopersicum*: 1 handful of leaves	All ingredients pounded and mixed with palm oil and/or alcohol.	Divided into two parts: One to rub the belly and the back, the other rolled into a ball introduced into the vagina when going to bed, once a day	[56]
Diabetes	Kenya	*Spilanthes mauritiana*	Not provided	Whole plant	Decoction	Oral	[57]
Diabetes	Nkoung–Khi Division West Region Cameroon	*Spilanthes africana* L.	Not provided	Whole plant	Decoction of 100 g	drink a lukewarm glass two times per day, morning and evening	[23]

7.4.10 Miscellaneous Conditions

Poisoning and snakebites are treated with *Acmella caulirhiza* in Cameroon, Nigeria, and the Democratic Republic of Congo. It is used to aid bone regeneration in Cameroon, treat rheumatism in Central Africa, fever in Madagascar, and to ward off evil spirits in Cameroon (Table 7.11).

7.5 COMPOUNDS ISOLATED FROM *ACMELLA CAULIRHIZA* DELILE

Three alkylamides, a steroid, and a fatty acid (Figure 7.2) have been isolated from *Acmella caulirhiza* Delile. A reducing sugar was also isolated, but the structure has not been published [29].

7.5.1 Alkylamides

Spilanthol (affinin), (2*E*, 6*Z*, 8*E*)-*N*-(2-methylpropyl)-2,6,8-decatrienamide) (**I**) and (2*E*, 4*E*)-*N*-(2-methylpropyl)-2,4-nonadien-8-ynamide (**II**) were extracted from the leaves of *Acmella caulirhiza* Delile by maceration of 611 g of dried leaves with 1 L hexane at room temperature. The resulting 5.1 g of the extract upon chromatography over silica gel yielded 5 mg of (**I**) from the 80% hexane in ethyl acetate system. 4.0 mg of (**II**) was obtained from a 70% hexane-30% ethyl acetate-mix [59].

Compound **III** (2*E*, 4*E*, 8*E*, 10*Z*)-*N*-(2-methylpropyl)-2,4,8,10-tetraenamide was isolated from a methanol extract of the fresh vegetative aerial parts of *Spilanthes mauritiana* [24]. The amide found was later identified as the 8*Z*, 10*Z* isomer rather than the 8*E*, 10*Z* [60]. Table 7.12 compares the ¹HNMR data obtained by Jondiko and Matovic for compound (**III**). The signals at 6.26 ppm [24] and 5.95 ppm [60] indicate the discrepancies between the different isomers.

7.5.2 Steroid and Fatty Acid

β-Stigmasterol (**IV**) and palmitic acid (**V**) were isolated from a pooled hexane and methylene chloride extract of fresh, whole *Spilanthes filicaulis* plants. These compounds were obtained from fractions corresponding to elution of a silica column of the extract with 100%–95% hexane [61].

7.6 INVESTIGATED BIOACTIVITY OF EXTRACTS AND ISOLATES

Members of the *Acmella* genus share several ethnomedicinal uses in common and may suggest several similar compounds present in the various species. In this section, extracts of *Acmella caulirhiza* will be discussed using the synonyms, as reported in the literature.

7.6.1 Analgesia, Hyperalgesia, and Antinociception

The dichloromethane extract of the whole plant of *Spilanthes africana* is weakly active at Mu opioid receptors in the brain. With an IC$_{50}$ of 98 μg/mL of action on opioid receptors, it may contribute to the plant's overall analgesic efficacy [62].

TABLE 7.11
Miscellaneous Conditions

Disease	People and Place	Plant Name	Alternative Names	Plant Part	Preparation	Administration	Reference
Poisoning	South-West Region of Cameroon	*Spilanthes filicaulis*	Not provided	Whole plant	Decoction	Not provided	[25]
Poisoning and antivenom	Ebiraland of Kogi State, Nigeria	*Spilanthes filicaulis* Schumach & Thorn	Osote	Leaves and stems	Macerated in water	Drink *ad libitum* or wash the affected body part with it	[51]
Anti-snake venom	Democratic Republic of Congo	*Spilanthes filicaulis* Schumach & Thorn	Not provided	Young leaves	Decoction	Not provided	[51]
Broken teeth and alveolar bone.	Cameroon	*Spilanthes africana*	Not provided	Not provided	Not provided	Applied directly to the cavity	[34]
Bone fractures	Cameroon	*Spilanthes Africana* L	Bangàa toue	Aerial parts	Maceration in cold water	Applied to fracture	[42]
Rajoo (description not provided)	Bale Mountains National Park, South Eastern Ethiopia	*Acmella caulirhiza* Del.	Not provided	The whole plant, fresh or dried	Concocted, crushed, powdered, and mixed with coffee and sugar.	Not provided	[43]
Rheumatism	Central Africa	*Spilanthes filicaulis*	Not provided	Leaves	Decoction	Wash affected body part with the decoction	[51]

(Continued)

TABLE 7.11 (*Continued*)
Miscellaneous Conditions

Disease	People and Place	Plant Name	Alternative Names	Plant Part	Preparation	Administration	Reference
Fever	Madagascar	*Acmella caulirhiza* Del	Acmelle	Whole plant	Not provided	Not provided	[33]
Ward off evil spirit	Menoua division, West Cameroon	*Spilanthes filicaulis* (S. &T.) Adams.	Pantouch	Whole Plant	Macerated in water with *Ageratum conyzoïde, Carica papaya, Mangifera indica, sidium guajava, Aframomum letestuanum, Ageratum conyzoïdes, Bidens Pilosa*	Two glasses per day	[54]
Swelling	Tara-gedam and Amba remnant forests of Libo Kemkem District, northwest Ethiopia	*Acmella caulirhiza* Del	. Kutchamelk	Leaves	Crush and powder	Humans and livestock: tie with honey/ butter	[58]

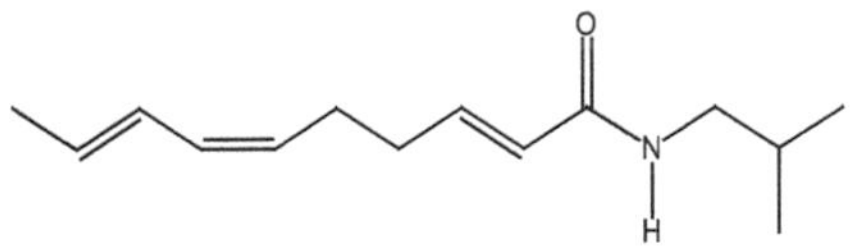

(2E,6Z,8E)-N-(2-methylpropyl)-deca-2,6,8-trienamide **(I)**

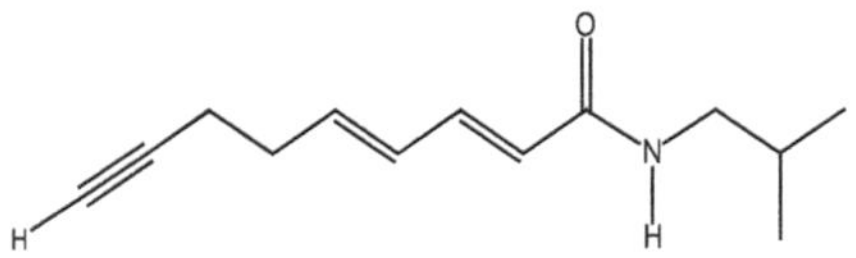

(2E,4E)-*N*-(2-methylpropyl)-2,4-nonadien-8-ynamide **(II)**

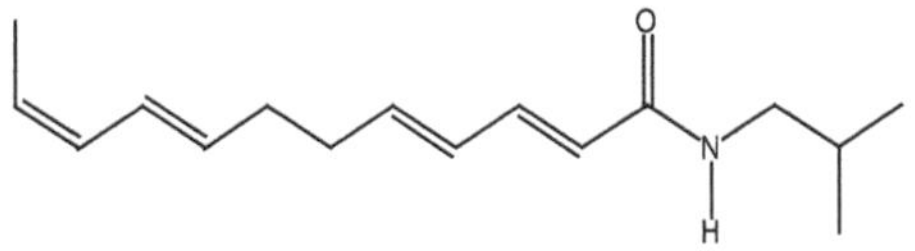

(2E,4E,8E,10Z)-N-(2-methylpropyl)-dodeca-2,4,8,10-tetraenamide **(III)**

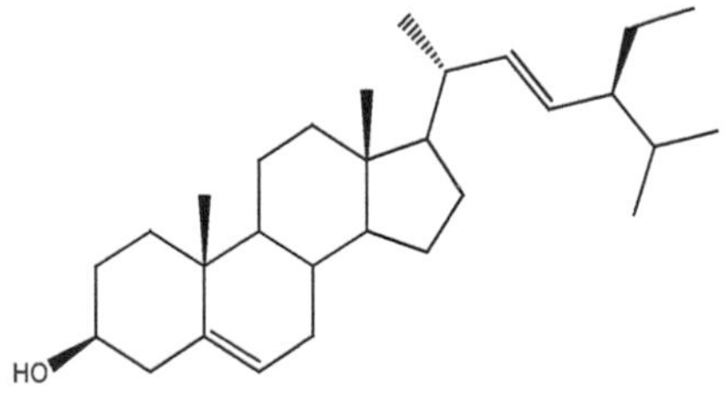

β-Stigmasterol **(IV)**

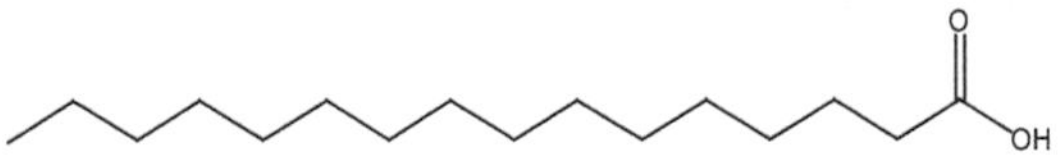

Palmitic acid **(V)**

FIGURE 7.2   Isolated compounds from *Acmella caulirhiza*.

An aqueous extract of *Spilanthes filicaulis* demonstrated a dose-dependent analgesic activity of 100–400 mg/kg. This study indicated that *Spilanthes filicaulis* could
be used to treat pain [63].

The analgesic activity of spilanthol has been attributed to an increased gamma-
aminobutyric acid (GABA) release in the temporal cerebral cortex [64].

TABLE 7.12
^{1}H NMR Data in CDCl$_3$ for Compound (3) [25,61]

[60]		[24]	
0.92	d, 6H, J = 6.8Hz	0.83	d, H-3′, H-4′, 6H, J = 7Hz
1.73	dd,3H, J = 2.0, 6.8 Hz	1.75	d, H-12, 3H, J = 7Hz
1.78	sept, 1H, J = 6.8 Hz	1.80	m, H-2′, 1H
2.20–2.30	m, 4H	2.25	m, H-6, H-7, 4H
3.15	dd, 2H, J = 6.8, 6.8 Hz	3.14	t, H-1′, 2H, J = 7Hz
5.39	dq, 1H, J = 7.2, 10.8 Hz	5.45	dq. H-IO, J = 10, 7 Hz
		5.60	br s, NH
5.57–5.66	m, 2H	5.62	dd, H-10, J = 10, 7Hz
5.77	d, 1H, J = 14.8Hz	5.76	d, H-2, 1H, J = 15Hz
5.95	dd,1H, J = 11.0, 11.5Hz		
6.05	dt,1H, J = 6.7, 14.8 Hz	6.07	dt, H-5, 1H, J = 15Hz
6.15	dd, 1H, J = 10.8, 15.2Hz	6.14	dd. H-4. 1H, J = 15, 10Hz
		6.26	dd, H-9, J = 15, 10Hz
6.34	dd,1H, J = 10.8, 15.2 Hz	6.33	dt, H-8, 1H, J = 15,7Hz
7.18	dd, 1H, J = 10.4, 14.8Hz	7.15	dd, H-3, 1H, J = 15, 10Hz

7.6.2 ANTIBACTERIAL ACTIVITY

Crude methylene chloride and hexane extracts of fresh, whole *Spilanthes filicaulis* plants demonstrated activity against several multi-drug resistant bacteria species isolated from patients with urinary tract infections. The organisms were *Escherichia coli, Klebsiella oxytoca, Enterobacter cloacae, Pseudomonas aeruginosa, Proteus vulgaris, Serratia marcescens, Citrobacter freundii, Klebsiella pneumoniae, Staphylococcus aureus*, and *Streptococcus species*. The most potent fraction from the methylene chloride extract had a M.I.C. of 0.06 mg/mL in contrast to that of the Ciprofloxacin standard with a M.I.C between 0.008 and 0.016 mg/mL.

β-Stigmasterol (**IV**) and palmitic acid (**V**) were isolated from nonpolar fractions of the pooled hexane and methylene chloride extracts. Both compounds had a M.I.C. of 0.1 mg/mL against *Salmonella typhimurium*. Also, stigmasterol had a M.I.C. of 0.1 mg/mL against *Escherichia coli*. β-stigmasterol had a M.I.C. of 0.2 mg/mL against *Salmonella typhi, Enterococcus faecalis, Klebsiella pneumoniae*, and *Proteus mirabilis* [61].

The bacteriostatic and bactericidal properties of a methanolic extract of the roots and flowers *Spilanthes mauritiana* from Kenya were investigated against 105 strains of bacteria. The M.I.C.s and M.B.C.s of ≥ 8 mg/mL^{-1} indicated a very low antibacterial activity. The exact values against the different bacterial strains were not reported for *Spilanthes mauritiana* [65]. These values validate the traditional use of this plant species for various infectious diseases. [59].

Three chloroform extracts of the flower heads, leaves, and stems of *Acmella caulirhiza* collected in Kericho County, Kenya, were tested against *Escherichia Coli, Staphylococcus aureus, Candida albicans*, and *Bacillus pumilus*. The stems had the highest antibacterial activity, but no antifungal activity was observed for all three extracts. These results further support the plant's ethnomedicinal use for the treatment of ulcers and sore throats [66].

A methanolic extract of *Spilanthes filicaulis* showed moderate M.I.C. = 64–512 µg/mL antimicrobial activity toward at least six *Staphylococcus aureus* clinical isolates [47].

An aqueous extract of *Spilanthes filicaulis* inhibited the growth of *Escherichia coli* and *Klebsiella pneumonia*. The L_{50} was 6.92 g/kg in mice after oral administration. It indicated that it could be used as an anti-infective agent [63].

Acmella caulirhiza Delile is utilized in the management of dental caries in rural communities of Uganda. Aqueous, methanol, and diethyl ether extracts showed antibacterial activity against *Streptococcus mutans* (ATCC 6519), a cariogenic bacterium in dental caries. The diethyl ether extract had the highest activity. The aqueous extract had the lowest MBC value of 250 mg/mL [67].

7.6.3 Bone Regeneration

Spilanthes africana twigs were collected in the Dschang region (Cameroon) to make an ethanolic extract. The extract's ability to regenerate bone was quantified by a calcein label (mineral deposition) at the site of a drill hole in a bone. The extract increased mineral deposition at 250 mg/kg (54% increase over control; $p < 0.05$), 500 mg/kg (60% increase over control; $p < 0.05$), and 750 mg/kg (131% over control; $p < 0.01$) doses [68]. These results are congruent with the ethnopharmacological use for fontanel closure in newborn babies in Gabon [53] and bone fracture treatment in Cameroon [42].

7.6.4 Antiviral Activity

The polar fraction of an ethanol extract of the leaves of *Acmella caulirhiza* Delile showed small to moderate antiviral action against herpes simplex virus type 1 and coxsackie B2 virus and an EC50 (µg/mL) of >140.72 against HIV [69].

7.6.5 Fungicidal Activity

A whole plant methanolic extract of *Acmella caulirhiza* Delile failed to show significant activity against Mycobacteria species [70].

Both the methanolic extracts of the roots and the flowers of *Spilanthes mauritiana* collected in Kenya displayed fungicidal activity against *Aspergillus niger*, *Aspergillus flavus*, and *Aspergillus fumigatus* (ATCC 90906, NCPF 2129 and NCPF 2140) [71].

An ethanolic leaf extract of *Spilanthes mauritiana* (Rich. et Pers.) D.C. collected in Rwanda did not display significant antifungal activity [69].

7.6.6 Antimalarial Activity

Methylene chloride, a 1:1 methylene chloride/methanol mixture, methanol, and water extracts of the stems of *Acmella caulirhiza* Delile were tested against chloroquine-sensitive *Plasmodium falciparum* D10. Their respective IC_{50} values were 38, 5.3, 64, and >100 µg/mL and did not indicate promising (IC_{50}< 10 µg/mL) or highly active (IC_{50}< 5 µg/mL) antimalarial activities [72].

7.6.7 INSECTICIDAL ACTIVITY

Aedes aegyptii mosquitoes are particularly crucial as they are vectors for dengue, yellow fever, and Zika virus. Yellow fever outbreaks were effectively controlled in the Americas by vector control programs, mainly targeting *Aedes aegyptii* in the 1950s and 1960s [73]. Dichlorodiphenyltrichloroethane (DDT) and the pyrethroids have been historically effective in controlling *Aedes aegyptii*, among other insect species. However, due to these insecticides' widespread use, the incidence of resistance to these agents has grown. Thus, there has been a concerted effort to find novel insecticides, focusing on natural insecticidal compounds [74]. Several members of the *Acmella* genus have been investigated for their insecticidal properties, whereas several extracts of *Spilanthes acmella* have been investigated for larvicidal properties. Spilanthol, a compound also isolated from *Acmella caulirhiza*, was extracted from the flower buds of *Spilanthes acmella* and found to have an LD_{50} of 4.5 ppm in third instar and 5 ppm in fourth instar *Aedes aegyptii* larvae [75]. In a similar experiment, three isobutyl amides, spilanthol, undeca-2E, 7Z, 9E-trienoic acid isobutyl amide, and undeca-2E-en-8,10-diynoic acid isobutyl amide was extracted from flowers of the same plant with hexane instead of petroleum ether. They are effective against fourth instar *Aedes aegyptii* larvae and caused 100% mortality at 12.5 µg/mL [76].

The LC_{50} of spilanthol, isolated from the hexane and methanol extracts of *Spilanthes acmella*, against *Plutella xylostella*, a moth species, reached 1.47 g/L. [77].

The ethyl acetate extract of *Spilanthes mauritiana* is effective against three mosquito species than the petroleum ether extract, with LC_{50} values of 14.10, 11.51, and 5.103 ppm, respectively, against adult members of *Anopheles stephensi*, *Aedes aegypti*, and *Culex quinquefasciatus*. At 100 ppm, the *Acmella caulirhiza* Delile extract inhibited the hatching of 100% of all three mosquito species' eggs. In addition to the ovicidal activity, *Acmella caulirhiza* Delile extract also deterred adult mosquitoes from laying their eggs where the extract was present in significant concentrations [78]. A similar study [24] on larvae of *Aedes aegyptii* showed that the chloroform extract of the leaves of *Acmella caulirhiza* Delile collected in the Kisii Highlands in Kenya was 100 times more potent than the methanolic extract. While both extracts showed mosquito larvicidal activity, the aqueous extract failed to show such activity. 100% mortality of the larvae occurred at 0.05 mg/mL of the methanol extract, while a similar complete killing was observed at concentrations as low as 0.0001 mg/mL of the chloroform extract. N-isobutyl-2E, 4E, 8E, 10Z-dodeca-2,4,8,10-tetraenamide was isolated from the chloroform extract and showed 100% mortality of the larvae at 10^{-5} mg/mL [33].

Research shows that alkylamides are responsible for the insecticidal, larvicidal, ovicidal, and oviposition of the *Acmella* genus and related genera rich in these amides. Various N alkylamides were synthesized and tested against superkdr-resistant house flies. This resistance mechanism increases the knockdown time of houseflies and decreases the overall efficacy (number of kills) of DDT and the pyrethroids. The alkylamides, similarly to DDT and the pyrethroids, act on sodium channels [79]. However, no cross-resistance is seen between both these classes of insecticidal compounds. On the contrary, some alkylamides are less sensitive to the resistance conferred by the super-kdr mechanism in houseflies, and thus provide valuable alternatives in areas where DDT resistance, particularly by the super-kdr mechanism, is prevalent [80].

7.6.8 ACUTE TOXICITY

The acute toxicity and sub-acute toxicity of the water extract of *Spilanthes africana* Delile were investigated in Wistar rats. Acute toxicity was evaluated by monitoring the behavioral effects of a single large oral dose of the extract at either 2000 or 5000 mg/kg in the test rats. In all the rats administered with the extract, a decrease in sensitivity and mobility was observed but no mortality. These changes were reversed entirely 24 h after the administration of the extract. The LD_{50} for a single oral dose is thus purported to be >5000 mg/kg body weight. Sub-acute toxicity was determined by evaluating the changes in rats' various systems and organs after daily oral administration of doses between 250 and 1000 g/kg bodyweight for 28 days. Some of the observed changes include an increase in the percentage of lymphocytes, while there was a significant decrease in monocytes' percentage. In male rats, throughout the treatment duration, triglyceride levels dropped and later increased after the treatment with the extract treatment was stopped. An increase in catalase and glutathione levels in female kidney tissue suggests that the extract may stimulate the production of free radicals and reactive oxygen species that contribute to the plant's toxicity. The investigators recommended that an oral dose <500 mg/kg body weight should be considered as the maximum daily oral dose to lessen the incidence of adverse reactions, reversible or otherwise [81].

7.6.9 ANTI-ULCER ACTIVITY

An aqueous extract of whole *Spilanthes filicaulis* plants collected in Yaounde and used to relieve stomach complaints in Cameroon was tested for anti-ulcer activity in ulcer-induced male Wistar rats. The extract produced complete mucosal cytoprotection at a dose of 2000 mg/kg, congruent with its ethnomedicinal use [45].

7.6.10 ANTIDIABETIC ACTIVITY

A 1:1:1 mixture of *Spilanthes africana* D.C., *Portulaca oleracea* Linx et, and *Sida rhombifolia* Linx extracts was orally administered to streptozotocin-induced diabetic rats in doses of 50, 100, and 200 mg/kg of body weight. Nondiabetic and diabetic control received water. The mixture of extracts decreased the levels of glycemia, lipid profile, total protein, malondialdehyde (MDA), glutathione, as well as aspartate aminotransferase (ASAT), alanine aminotransferase (ALAT), and creatinine significantly ($p < 0.05$) in a dose-dependent manner. The mixture also displayed DPPH and hydroxy-radical scavenging. The mixture of extracts also increased the concentration of HDL cholesterol, total protein, glutathione, and the total antioxidant status in the treated group. The mixture of extracts also had a scavenging effect on DPPH and OH radicals. These results suggest that the mixture of extracts could be used to treat diabetes mellitus [82].

7.7 BIOSYNTHESIS OF ALKYLAMIDES

Plants of the Asteraceae family are known for accumulating various types of alkylamides in their stems, flowers, leaves, and roots. Several of these alkylamides can be found across various genera of this family. However, the alkylamide profile of any given genus may both serve to distinguish it from other genera of the class and reflect evolutionary trends between closely associated genera. It is believed that these secondary plant metabolites are synthesized by the plant to play a role in plant regeneration, particularly adventitious root formation through a nitric oxide-dependent mechanism. Spilanthol, a compound common to several genera of the Asteraceae family, and an N-isobutyl decanamide were used to establish this function of alkylamides in plants [83]. In a study with *Acmella radicans*, the biosynthesis of spilanthol was observed as early as the end of the first week after seed germination. Levels of spilanthol rose between the third and fifth weeks of germination, with the highest concentrations of spilanthol and other alkylamides generated between the ninth and tenth weeks of seed germination. Contrary to studies done in *Echinacea*, where the highest alkylamide concentration was found in roots, flower heads of *Acmella radicans* accumulated the most alkylamides [84].

Various alkylamides synthesized in various plants possess very different structures. Alkylamides consist of two main parts: the acid part and the amine part (Figure 7.3).

The acid parts could be aliphatic saturated, aliphatic olefinic, and aliphatic acetylenic. They can also, for example, possess an aromatic ring or a piperidine residue. The amine parts could be N-isobutyl, N-2-methyl butyl, 2-phenyl ethyl, and cyclic amines such as piperidinyl and pyrrolidinyl. Aliphatic amides are the most commonly found amides in the Asteraceae family; as such, the *Acmella* and *Spilanthes* genera have mostly aliphatic amides with 8–12 carbon atom residues in the acid parts and predominantly isobutyl and methyl butyl amine residues [64] (Figure 7.4).

The biosynthesis of these amides in various plants follows a similar general pattern. Coenzyme A binds to malonic acid residues to form fatty acid chains of growing

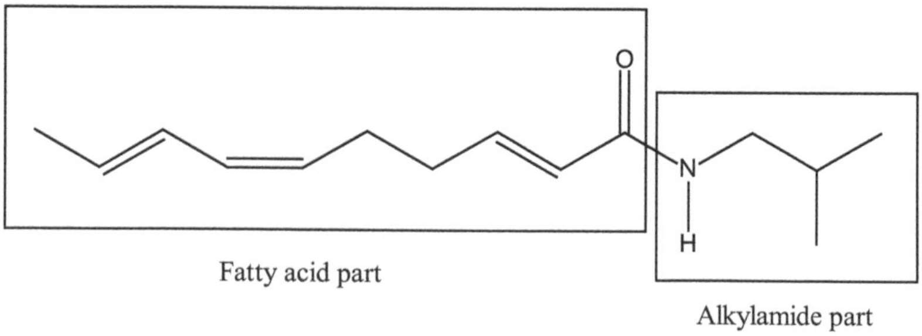

Fatty acid part

Alkylamide part

FIGURE 7.3 Alkylamide structure.

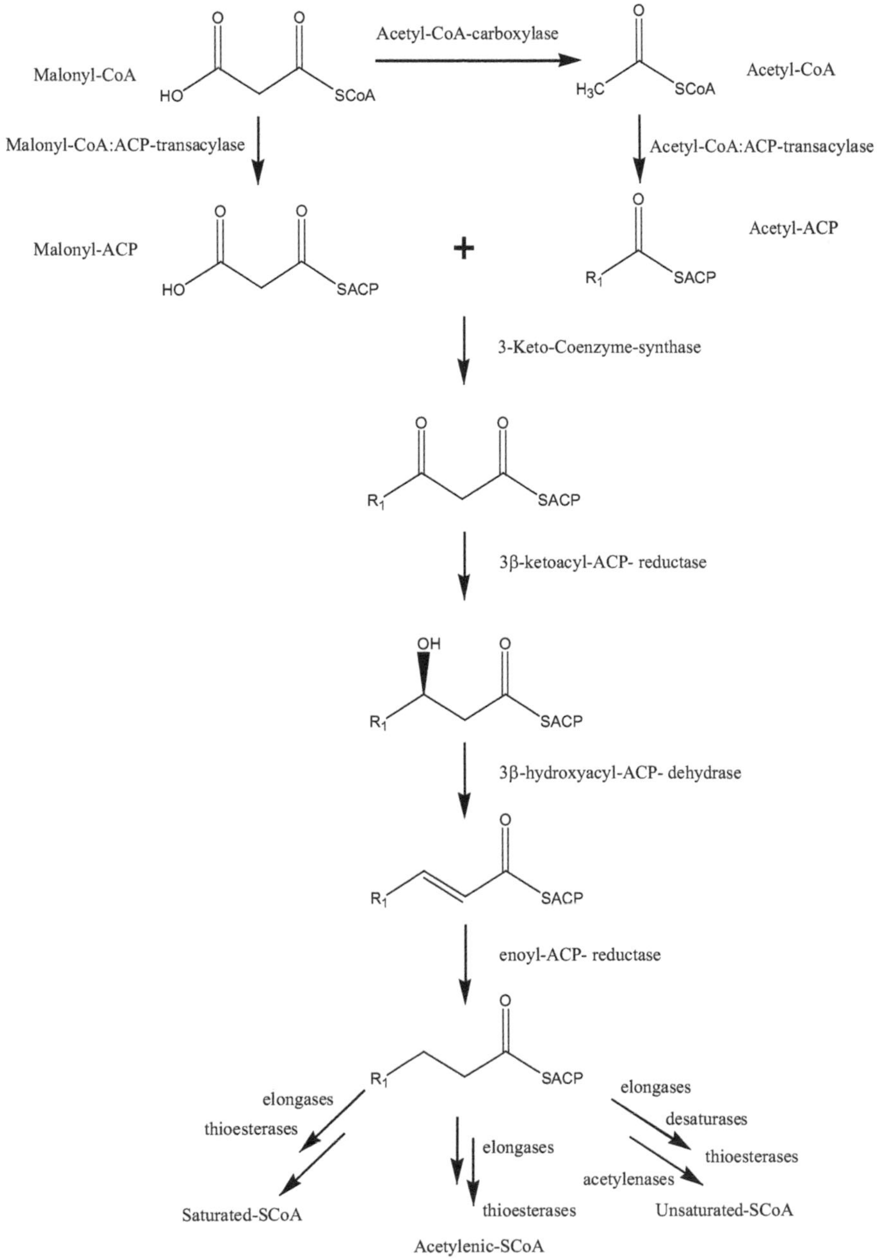

FIGURE 7.4 The fatty acid pathway in the biosynthesis of alkylamides.

lengths catalyzed by fatty acid synthases. The addition of carrier-bound malonyl units occurs in a head-to-tail direction. The acyl chain grows by two carbon atoms upon each addition of a malonyl unit. This process is a combination of Claisen condensation reactions, reduction to a hydroxy acyl, dehydration, and subsequent reduction of the unsaturated acyl chain to form a saturated acyl chain two carbons longer. The enzyme system that catalyzes this reaction is often referred to as the elongase system. It consists of four enzymes that catalyze each of the above reactions. The β-ketoacyl CoA synthase appends malonyl coenzyme A molecules to an existing acetyl co A – carrier protein complex. The β-ketoacyl reductase then catalyzes the formation of the β-hydroxy ester. Subsequent dehydration by β-hydroxy acyl CoA dehydrase yields a trans α, β unsaturated ester in a stereospecific manner. This unsaturation is subsequently reduced to a saturated acyl chain via trans-2-enoyl reductase [85].

Stearic acid and its C9 monounsaturated derivative oleic acid form the basis for synthesizing the acid parts of most aliphatic alkylamides in plants. It occurs through a combination of chain elongation or shortening processes, desaturation, reduction, and dehydrogenation. The desaturation process usually involves regiospecific desaturases such as the stearoyl acyl protein desaturase, which inserts a double bond on the stearic acid molecule between C9 and C10 of the unsaturated chain [85]. Further action by relevant desaturases may convert the oleic acid obtained to linoleic acid (Δ^{12} desaturase) and, subsequently, linolenic acid (Δ^{15} desaturase) as dictated by the target product [86]. The formation of triple bonds is catalyzed by acetylenes, which are regiospecific. The acetylenic amides of *Acmella* are believed to be derivatives of the crepenynic acid form of linoleic acid by the action of Δ^{12} acetylenase [87]. For acyl chains longer than 18C, elongation of the fatty acid functions in the same way as previously mentioned through the elongase system. Amides found in this plant family are usually of carbon chain length between 8 and 12 C atoms. Thus, chain shortening processes are necessary for the biosynthetic pathway. It is often achieved through beta-oxidation of the corresponding longer chain unsaturated fatty acid coenzyme A complexes (Figure 7.5).

Alkylamides of the *Acmella* and *Spilanthes* genera have three ubiquitous amide units: N-isobutyl amides, 2-methyl butyl amides, and 2-phenyl ethyl amides. The precursors for these amides are valine, isoleucine, and phenylalanine, respectively. They are appended to the acid moieties via decarboxylation or possibly through other transformation processes [88].

7.8   CHEMICAL SYNTHESIS OF ALKYLAMIDES

At least four research groups have published total syntheses of spilanthol, one of the major metabolites isolated from *Acmella caulirhiza* and other plants in the

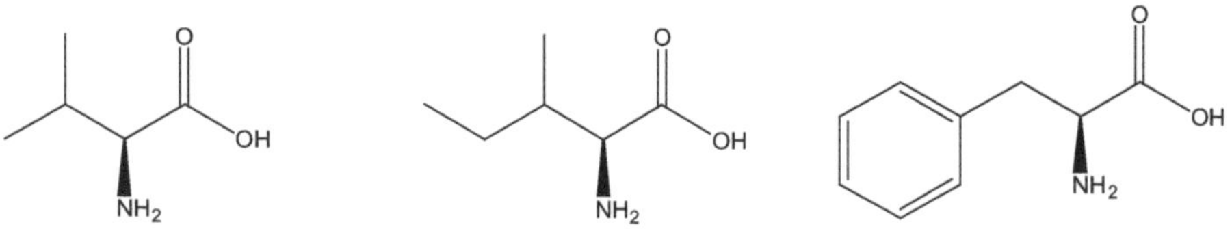

Valine   Isoleucine   Phenylalanine

FIGURE 7.5   Amino acid precursors of alkylamides in the *Acmella* and *Spilanthes* genera.

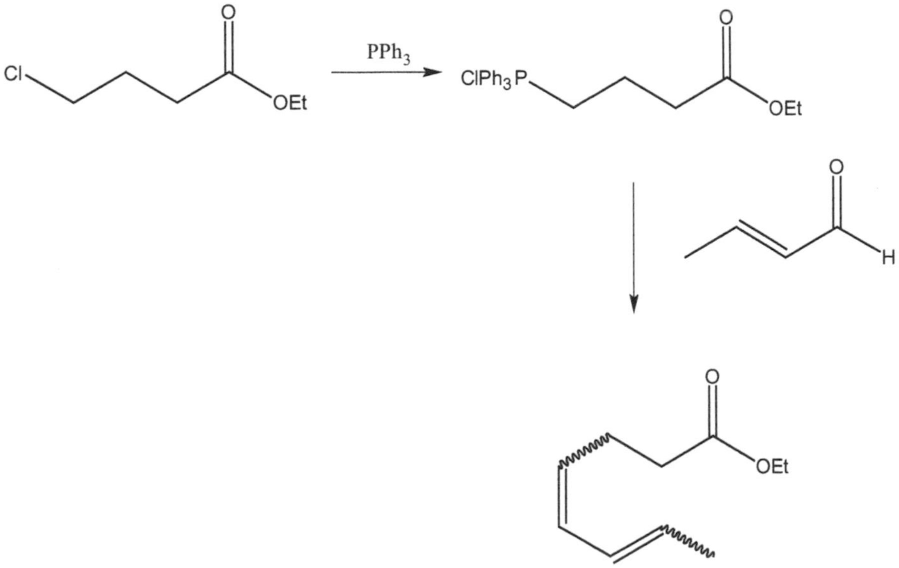

FIGURE 7.6   The Wittig reaction in the Takasago Synthesis. (Adapted from [94])

same and closely related genera [89–92]. Also, several patents have been filed by Takasago International Corporation [93,94]. One of the main aspects of these syntheses is the formation of allylic 6,8-diene bonds focusing on the stereochemistry as 6Z, 8E (Figure 7.6).

The Sonogashira reaction was used to form an acetylenic bond, followed by a stereoselective reduction to introduce the Z, E diene system [90]. The Sonogashira reaction involves the coupling of pent-4-yn-1-ol to a 60:40 E:Z racemic mixture of 1-bromo-1-propene over a palladium copper catalyst system. The product of the Sonogashira reaction can then be reduced to form the Z or E diene. The method used to achieve an almost complete Z selectivity of 95:5 in the semi reduction of alkynes involves the addition of trimethylsilyl chloride as a modification to Boland's method, which uses a Zn (Ag/Cu) catalyst alloy in the presence of water and methanol [95]. Strides have been made in the stereoselective semi-hydrogenation of alkynes to alkenes. It is possible to target and obtain the E or Z isomer in sufficiently high purity based on the catalyst system used or sometimes just by altering reaction conditions using the same catalyst [96] (Figure 7.7).

The formation of the 2E bond in spilanthol and other alkylamides often occurs during the amidation process. In Yamamoto's spilanthol synthesis, a Wittig reaction was used with the reagent formed from isobutyl triphenyl phosphonium acetamide reacted with the carbonyl group of (Z, E)-4,6-octadienal. It produced spilanthol or N isobutyldecatri-2E, 6Z, 8E-enamide [89], thereby introducing the final olefin bond and the amide moiety. The Horner–Wadsworth–Emmons reaction [97] was used to attach the isobutyl amide moiety to the octadienal intermediate used previously. The stabilized phosphonate carbanion produced the desired E-alkene with a >95:5 stereoselectivity for the E over the Z-conformer. However, in a patented spilanthol

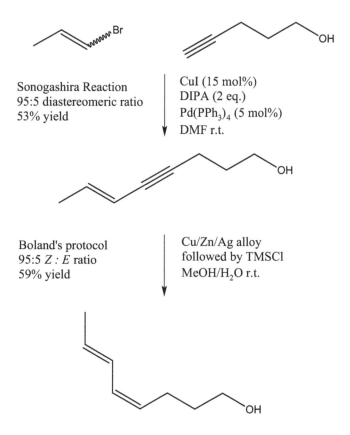

FIGURE 7.7 Stereoselective reduction of an alkyne. (Adapted from [90])

synthesis [94], the isobutyl amide is formed by substituting the 3-hydroxy methyl ester of spilanthol. The 3-hydroxy N-isobutyl amide is then converted to the 3-ester derivative, which can be reacted with a strong base such as an alkoxide to produce the 2E olefin bond (Figure 7.8).

7.9 MECHANISM OF ACTION OF SIGNIFICANT BIOACTIVITY

7.9.1 PLANT ALKYLAMIDES AND THE CENTRAL NERVOUS SYSTEM

Alkylamides, such as spilanthol, [98] have been shown to cross the skin in an experiment using Franz diffusion cells with human skin. It would support its use as a topical anesthetic or anti-inflammatory, as has been previously reported. Spilanthol and pellitorine (deca-2E, 4E-dienoic acid isobutyl amide) are alkylamides prevalent in the *Acmella* and *Echinacea* genera, respectively. These two N-alkylamides, among others, have been shown to penetrate the stratum corneum and make their way into general circulation in the dermis [99]. Spilanthol was also shown to be systemically present when applied to the buccal mucosa of pigs. Caco-2 monolayers have been used as in vitro methods with comparable results in vivo for the passive intestinal absorption of pharmacologically active molecules [100]. In a study of various alkylamides

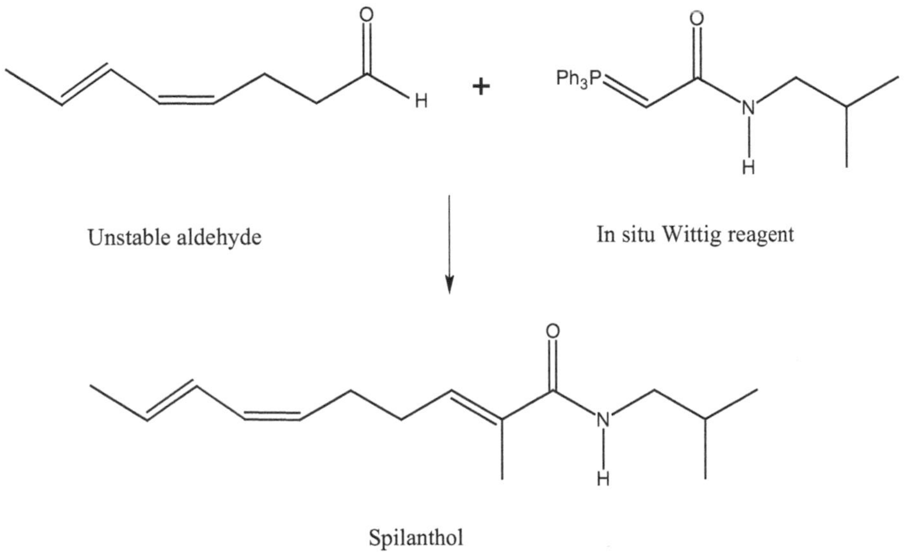

FIGURE 7.8 Yamamoto's Wittig reaction. (Adapted from [89])

associated with the *Echinacea* genus, their permeability through a Caco-2 cell mono-layer was evaluated and compared among each other for structure-based trends in their permeabilities. The membrane was generally more permeable to 2,4 dienes compared to their 2-ene alkylamide equivalents. Another trend observed [101] was that an increase in unsaturation seemingly increased the apparent permeability of the alkylamides. Among these alkylamides tested, Dodeca-2*E*, 4*E*, 8*E*, 10*E*/*Z*-tetraenoic acid isobutyl amides are common in the *Echinaceae* genera. They have also been iso-lated from *Acmella caulirhiza* Delile and *Spilanthes acmella* [102]. Pharmacokinetic and pharmacodynamic studies of the absorption and distribution of these tetraenoic acid amides from an oral dose have been carried out in rats. Within 15 min of admin-istration, this amide is present in significant concentrations in the brain, liver, and plasma. Most are found in the brain, mainly the striatum, hippocampus, cerebral cortex, and cerebellum, with decreasing concentrations in that order. A longer half-life and a lower clearance rate were observed in brain tissue than in plasma, with the amide lasting up to 6 h in some areas of the brain. The maximum concentration of alkylamide detected 15 min after an oral dose of 2.5 mg/kg in plasma, liver tis-sue, striatum, hippocampus, cerebral cortex, and cerebellum are, respectively, 26.35, 54.65, 45.96, 39.56, 36.49, and 33.80 ng/g [103].

7.9.2 CANNABINOID RECEPTOR INTERACTIONS

The endogenous cannabinoid system consists of at least two receptor types that have been cloned and expressed in various cells. Its endogenous ligands are generally small lipid molecules structurally similar to arachidonic acid derivatives. The major endo-cannabinoids are 2-arachidonoylglycerol (2-AG) and N-arachidonoylethanolamide or anandamide. Anandamide is a partial agonist at both cannabinoid receptors, while 2-AG is a full agonist. Other endogenous compounds with putative activity at the cannabinoid receptors are O-arachidonoylethanolamine (virodhamine), 2-arachidonoylglycerol ether (noladin ether), and arachidonoyl dopamine. Also associated with the endo-cannabinoid system are fatty acid amides such as the palmitoyl-, stearoyl-, and oleoyl- ethanolamides. These amides and their unsaturated isomers do not directly affect the cannabinoid receptors but through some yet unknown method of action enhance the cannabinoid action of the major endocannabinoids. These are often referred to as "entourage compounds" [104]. The cannabinoid receptor type 1, CB1, was the first to be cloned. It is found mostly in the CNS while the cannabinoid receptor 2 is more often associated with the periphery, particularly cells and tissues associated with immune response [105]. However, more recently, CB1 receptors have been shown to exist in the periphery, while CB2 receptors have been found in the CNS [106]. Both receptor subtypes 1 and 2 are G protein-coupled receptors. The effects of activation of these receptors by their agonists are mediated either by the inhibition of adenylyl cyclase or the activation of mitogen-activated protein kinases [107]. The activation of these receptors is often associated with a decrease in cyclic AMP, the activation of outward potassium currents, and the inhibition of calcium ion influx and, thus, neurotransmitter inhibition release [108, 109]. In addition to well-known phytocannabinoids, some natural products with possible cannabimi-metic activity include N-alkylamides like echinacein, triterpenes such as euphol, and sesquiterpenes like β-caryophyllene [109].

7.9.3   CANNABINOID RECEPTOR 1

Plant alkylamides interact with CB1 receptors. N-alkylamides extracted from *Heliopsis helianthoides* and N-benzyl amides from *Lepidium meyenii* were tested against several endocannabinoid targets such as anandamide cellular uptake, fatty acid amide hydro-lase, CB1Rs, and CB2Rs. Among the tested compounds, two of them, hexadeca-2*E*, 4*E*, 9,12-tetraenoic acid 2′-methyl butyl amide from *H. helianthoides* and N-benzyl-9*Z*, 12*Z*-octadecadienamide from *L. meyenii*, showed submicromolar (0.31 and 0.48 μM, respectively) binding affinities for CB1. They were both preferentially bound to CB1 than CB2 with their CB2 affinities in the low micromolar range (<5 μM). The ben-zyl amide also showed a submicromolar IC_{50} concentration for the inhibition of anandamide cellular uptake [110]. Conversely, the isobutyl amides from *Echinacea purpurea* showed inverse agonist action at CB1 receptors, while the methyl butyl amides showed a partial agonistic action. These amides' alkyl chains typically ranged from 11 to 13 carbon atoms with varying degrees of unsaturation [111]. Plant N-alkylamides are more likely to target CB2Rs.

7.9.4   CANNABINOID RECEPTOR 2

N-alkylamides have been shown in several studies to bind to and positively activate CB2Rs [60]. One study showed that dodeca-2*E*, 4*E*, 8*Z*, 10*Z*-tetraenoic acid isobu-tyl amide and dodeca-2*E*, 4*E*-dienoic acid isobutyl amide have a higher affinity for CB2 than endogenous ligand 2-AG. Like 2-AG, binding of these alkylamides to CB2 stimulates transient calcium ion currents in CB2-positive HL-60 cells and inhibits

TNF-α, IL-1β, IL-12p70, suggesting an anti-inflammatory role. In this study, undeca-2*E*-ene, 8,10-diynoic acid isobutyl amide, which has no significant affinity for CB1 receptors, showed some anti-inflammatory activity similar to that shown by the CB2 selective alkylamides and anandamide, suggesting the action of these alkylamides at a non CB1/CB2 receptor to which these alkylamides as well as anandamide show some affinity [112]. Several alkylamides (≥11 carbon atom alkyl chains) from the *Echinacea* genus have been shown to have an affinity for CB2Rs with their k_i values below 20 µM [113]. Synergistic induction of anti-inflammatory cytokine IL-6 and inhibition of TNF-α was observed in the alkylamide fractions of Echinacea extracts [114]. Some of these alkylamides are believed to exert their endocannabinoid system activation through the inhibition of FAAH (which breaks down endocannabinoids) and improving the transport of endocannabinoids through micellisation [105]. Like the well-documented entourage compounds such as palmitoylethanolamide, plant alkylamides might also potentiate the action of endocannabinoids, and thus their administration leads to indirect CBR activation [115]. However, the action of pal-mitoylethanolamide is not believed to be limited to inhibition of endocannabinoid metabolism but mediated by direct action at an as yet unidentified and uncharac-terized CB2-like receptor [116]. CB2 antagonists/inverse agonists have also shown promising antinociception and anti-inflammatory activity [117]. This could plausibly be another mode of action through which N-alkylamides exhibit anti-inflammatory effect through CB2 interaction.

REFERENCES

1. Moore, A.H., 1907. Revision of the genus spilanthes. *Proceedings of the American Academy of Arts and Sciences* 42(20): 521, Doi: 10.2307/20022250.
2. Jansen, R.K., 1981. Systematics of spilanthes (Compositae: Heliantheae). *Systematic Botany* 6(3): 231–257, Doi: 10.2307/2418284.
3. Mabberley, D., 2008. *Mabberley's plant book: A portable dictionary of plants, their classifications and uses*. 3rd ed., Cambridge: Cambridge University Press.
4. Roskov, Y., et al., 2013. Catalogue of life: Acmella caulirhiza Delile. Doi: ISSN 2405–8858.
5. GBIF Secretariat., n.d. Global biodiversity information facility. Doi: 10.15468/39omei.
6. Acmella caulirhiza Delile. The Plant List Version 1.1. http://www.theplantlist.org/tpl1.1/record/gcc-10941. Accessed: 20-Aug-2011.
7. African Plant Database., n.d. Acmella caulirhiza Delile. C.J.B. - African Plant Database. www.ville-ge.ch/musinfo/bd/cjb/africa/details.php?langue=an&id=104723. Accessed: 14-May-2020.
8. Jansen, R.K., 1985. The systematics of Acmella. *Systematic Botany Monographs* 8: 1–115.
9. Idu, M., Obaruyi, G.O. and Erhabor, J.O., 2008. Ethnobotanical uses of plants among the binis in the treatment of ophthalmic and E.N.T. (Ear, Nose and Throat) ailments. *Ethnobotanical Leaflets* 13: 480–496.
10. Kipkore, W., et al., 2014. A study of the medicinal plants used by the Marakwet Community in Kenya. *Journal of Ethnobiology and Ethnomedicine* 10(1), Doi: 10.1186/1746–4269–10–24.
11. Seshathri, K. and Thiyagarajan, T., 2011. Antimicrobial activity of chewing sticks of Jimma – Ethiopia against Streptococcus pyogens. *Journal of Phytology* 3(8): 34–37.

12. Cos, P., et al., 2002. C omplement modulating activity of Rwandan medicinal plants. *Phytomedicine* 9(1): 56–61, Doi: 10.1078/0944–7113–00085.

13. Amsalu, N., et al., 2018. Use and conservation of medicinal plants by indigenous people of Gozamin Wereda, East Gojjam Zone of Amhara Region, Ethiopia: An ethnobotanical approach. *Evidence-Based Complementary and Alternative Medicine*, Doi: 10.1155/2018/2973513.

14. Giday, M., Asfaw, Z. and Woldu, Z., 2010. Ethnomedicinal study of plants used by Sheko ethnic group of Ethiopia. *Journal of Ethnopharmacology* 132(1): 75–85, Doi: 10.1016/j.jep.2010.07.046.

15. Woldeamanuel, B.B., 2016. *A study on medicinal plants of Gamo People: The case of Bonke Woreda, SNNPR, Ethiopia*. Arba Minch University, Arba Minch.

16. Seid, M.A. and Aydagnehum, S.G., 2013. Medicinal plants biodiversity and local Healthcare management system in Chencha District; Gamo Gofa, Ethiopia. *Journal of Pharmacognosy and Phytochemistry* 2(1): 284–293.

17. Abera, B., 2014. Medicinal plants used in traditional medicine by Oromo people, Ghimbi district, Southwest Ethiopia. *Journal of Ethnobiology and Ethnomedicine* 10(40), Doi: 10.1186/1746–4269–10–40.

18. Atnafu, H., et al., 2018. Ethnobotanical study of medicinal plants in selale mountain-ridges, North Shoa, Ethiopia. Lundiana: *International Journal of Biodiversity* 2(6): 567–577.

19. Kidane, B., et al., 2014. Use and management of traditional medicinal plants by Maale and Ari ethnic communities in southern Ethiopia. *Journal of Ethnobiology and Ethnomedicine* 10: 46, Doi: 10.1186/1746–4269–10–46.

20. Ndjouondo, G.P., et al., 2015. Inventaire et caractérisation des plantes médicinales des sous bassins versants Kambo et Longmayagui (Douala, Cameroun). *Journal of Animal & Plant Sciences* 25(3): 3898–3916.

21. Ndenecho, E.N., 2009. Herbalism and resources for the development of ethnopharmacology in Mount Cameroon region. *African Journal of Pharmacy and Pharmacology* 3(3): 078–086.

22. Mathias, S.N., Ilyas, N. and Musa, K.Y., 2013. Ethnomedical study of plants used as therapy against some digestive system ailments among Takkad People of Kaduna State, Nigeria. *Nigerian Journal of Pharmaceutical Sciences* 12(1): 1–6.

23. Koyeu, T.E., et al., 2014. Ethnobotanic contribution of Cameroon: Anti-diabetic plants inventory in the Nkoung–Khi division West Region Cameroon. *Applied Science Reports* 8(3): 125–133, Doi: 10.15192/pscp.asr.2014.4.3.125133.

24. Jondiko, I.J.O., 1986. A mosquito larvicide in Spilanthes mauritiana. *Phytochemistry* 25(10): 2289–2290, Doi: 10.1016/S0031–9422(00)81681–7.

25. Jiofack, T., et al., 2009. Ethnobotanical uses of some plants of two ethnoecological regions of Cameroon. *African Journal of Pharmacy and Pharmacology* 3(13): 664–684.

26. Betti, J.L. and Lejoly, J., 2009. Contribution to the knowledge of medicinal plants of the Dja Biosphere Reserve, Cameroon: Plants used for treating jaundice. *Journal of Medicinal Plants Research* 3(12): 1056–1065.

27. Gebeyehu, G., 2011. *An ethnobotanical study of traditional use of medicinal plants and their conservation status in Mecha Wereda, West Gojam Zone of Amhara Region, Ethiopia*. Addis Ababa University, Addis Ababa.

28. Megersa, M., Jima, T.T. and Goro, K.K., 2019. The use of medicinal plants for the treatment of toothache in Ethiopia. *Evidence-Based Complementary and Alternative Medicine* 2019, Doi: 10.1155/2019/2645174.

29. Watt, J.M. and Breyer-Branderwijk, M.G., 1962. The medicinal and poisonous plants of southern and eastern Africa. 2nd ed., Edinburgh and London: E & S Livingstone L.T.D.

30. Okello, J. and Ssegawa, P., 2007. Medicinal plants used by communities of Ngai Subcounty, Apac District, northern Uganda. *African Journal of Ecology* 45(SUPPL. 1): 76–83, Doi: 10.1111/j.1365–2028.2007.00742.x.
31. Flatie, T., et al., 2009. Ethnomedical survey of Berta ethnic group Assosa Zone, Benishangul-Gumuz regional state, mid-west Ethiopia. *Journal of Ethnobiology and Ethnomedicine* 5(14), Doi: 10.1186/1746–4269–5–14.
32. Akendengue, B. and Louis, A.M., 1994. Medicinal plants used by the Masango people in Gabon. *Journal of Ethnopharmacology* 41: 193–200.
33. Gurib-Fakim, A., Gueho, J. and Sewraj-Bissoondoyal, M., 1997. The medicinal plants of Mauritius - Part 1. *Pharmaceutical Biology* 35(4): 237–254, Doi: 10.1076/ phbi.35.4.237.13313.
34. Agbor, A.M., 2015. Methanol extracts of medicinal plants used for oral healthcare in Cameroon. *Biochemistry & Pharmacology* 4(2): 5, Doi: 10.4172/2167–0501.1000164.
35. Eshetu, G.R., et al., 2015. Ethnoveterinary medicinal plants: Preparation and application methods by traditional healers in selected districts of southern Ethiopia. *Veterinary World* 8(5): 674–684, Doi: 10.14202/vetworld.2015.674–684.
36. Namukobe, J., et al., 2011. Traditional plants used for medicinal purposes by local communities around the Northern sector of Kibale National Park, Uganda. *Journal of Ethnopharmacology* 136(1): 236–245, Doi: 10.1016/j.jep.2011.04.044.
37. Noumi, E., 2015. Ethno-medico-botany, theory, and philosophy: case of traditional medicinal plant uses in hemiplegia and neuralgia, in Bui Division, northwest region of Cameroon. *World Journal of Pharmaceutical Research* 4(12): 377–392.
38. Noumi, E. and Fozi, F.L., 2003. Ethnomedical botany of epilepsy treatment in Fongo-Tongo village, Western Province, Cameroon. *Pharmaceutical Biology* 41(5): 330–339, Doi: 10.1076/phbi.41.5.330.15944.
39. Noumi, E. and Yomi, A., 2001. Medicinal plants used for intestinal diseases in Mbalmayo Region, Central Province, Cameroon. *Fitoterapia* 72(3): 246–54, Doi: 10.1016/S0367-326X(00)00288–4.
40. Tsobou, R., Mapongmetsem, P.M. and Van Damme, P., 2013. Medicinal plants used against typhoid fever in Bamboutos division, western Cameroon. *Ethnobotany Research and Applications* 11: 163–74, Doi: 10.17348/era.11.0.163–174.
41. Fenetahun, Y. and Eshetu, G., 2017. A review on ethnobotanical studies of medicinal plants use by agro-pastoral communities in Ethiopia. *Journal of Medicinal Plants Studies* 5(1): 33–44.
42. Hubert, D.J., et al., 2013. In vitro leishmanicidal activity of some Cameroonian medicinal plants. *Experimental Parasitology* 134(3): 304–308, Doi: 10.1016/j.exppara. 2013.03.023.
43. Yineger, H., et al., 2008. Plants used in traditional management of human ailments at BaleMountains National Park, Southeastern Ethiopia. *Journal of Medicinal Plants Research* 2(6): 132–153.
44. Isah, A.O., et al., 2015. Ethnomedical survey of some of the plants used for pain management in Lokoja, Kogi State, Nigeria. *Bayero Journal of Pure and Applied Sciences* 8(1): 72–79, Doi: 10.4314/bajopas.v.
45. Tan, P. V., Njimi, C.K. and Ayafor, J.F., 1997. Screening of some African medicinal plants for antiulcerogenic activity: Part 1. *Phytotherapy Research* 11(1): 45–47, Doi: 10.1002/(SICI)1099–1573(199702)11:1<45::AID-PTR945>3.0.CO;2–I.
46. Noumi, E. and Dibakto, T.W., 2000. Medicinal plants used for peptic ulcer in the Bangangte region, western Cameroon. *Fitoterapia* 71(4): 406–412, Doi: 10.1016/S0367-326X(00)00144–1.
47. Fonkeng, L.S., et al., 2015. Anti-Staphylococcus aureus activity of methanol extracts of 12 plants used in Cameroonian folk medicine. *Complementary and Alternative Medicine. BMC Research Notes* 8(1): 4–9, Doi: 10.1186/s13104-015-1663–1.

48. Jiofack, T., et al., 2009. Ethnobotany and phytomedicine of the upper Nyong valley forest in Cameroon. *African Journal of Pharmacy and Pharmacology* 3(4): 144–150.
49. Telefo, P.B., et al., 2011. Ethnopharmacological survey of plants used for the treatment of female infertility in Baham, Cameroon. *Journal of Ethnopharmacology* 136: 178–187, Doi: 10.1016/j.jep.2011.04.036.
50. Noumi, E., 2010. Ethno medicines used for treatment of prostatic disease in Foumban, Cameroon. *African Journal of Pharmacy and Pharmacology* 4(11): 793–805.
51. Atawodi, S., et al., 2014. Ethnomedical survey of Adavi and Ajaokuta local government areas of Ebiraland, Kogi State, Nigeria. *Annual Research & Review in Biology* 4(24): 4344–4360, Doi: 10.9734/arrb/2014/8658.
52. Yigezu, Y., Haile, D.B. and Ayen, W.Y., 2014. Ethnoveterinary medicines in four districts of Jimma zone, Ethiopia: Cross-sectional survey for plant species and mode of use. *BMC Veterinary Research* 10, Doi: 10.1186/1746–6148–10–76.
53. van Vliet, E., 2014. *Medicinal plants for women's en children's health in urban and rural areas of Gabon*. Utrecht University, Utrecht.
54. Yemele, M.D., et al., 2015. Ethnobotanical survey of medicinal plants used for pregnant women's health conditions in Menoua division-West Cameroon. *Journal of Ethnopharmacology* 160: 14–31, Doi: 10.1016/j.jep.2014.11.017.
55. Kamatenesi-Mugisha, M. and Oryem-Origa, H., 2007. Medicinal plants used to induce labour during childbirth in western Uganda. *Journal of Ethnopharmacology* 109(1): 1–9, Doi: 10.1016/j.jep.2006.06.011.
56. Noumi, E., 2010. Treating fibromyoma with herbal medicines in South Cameroon. *Indian Journal of Traditional Knowledge* 9(4): 736–741.
57. Kamau, L.N., et al., 2016. Knowledge and demand for medicinal plants used in the treatment and management of diabetes in Nyeri County, Kenya. *Journal of Ethnopharmacology* 189: 218–229, Doi: 10.1016/j.jep.2016.05.021.
58. Chekole, G., Asfaw, Z. and Kelbessa, E., 2015. Ethnobotanical study of medicinal plants in the environs of Tara-gedam and Amba remnant forests of Libo Kemkem District, northwest Ethiopia. *Journal of Ethnobiology and Ethnomedicine* 11: Doi: 10.1186/1746–4269–11–4.
59. Crouch, N.R., et al., 2005. A novel alkylamide from the leaves of Acmella caulirhiza (Asteraceae), a traditional surface analgesic. *South African Journal of Botany* 71(2): 228–230, Doi: 10.1016/S0254–6299(15)30137–X.
60. Matovic, N., et al., 2007. Stereoselective synthesis, natural occurrence and CB2 receptor binding affinities of alkylamides from herbal medicines such as Echinacea sp. *Organic and Biomolecular Chemistry* 5(1): 169–174, Doi: 10.1039/b615487e.
61. Akoachere, J.-F., et al., 2015. In vitro antimicrobial activity of agents from spilanthes filicaulis and Laportea ovalifolia against some drug resistant bacteria. *British Journal of Pharmaceutical Research* 6(2): 76–87, Doi: 10.9734/bjpr/2015/15582.
62. Efange, S. and Mash, D.C., 2004. *Drug development and conservation of biodiversity in west and central Africa: Performance of neurochemical and radio receptor assays of plant extracts drug discovery for the central nervous system*. Miami, FL: apps.dtic.mil.
63. Ekor, M., et al., 2005. Acute toxicity, analgesic potential and preliminary antimicrobial studies of the aqueous plant extract of spilanthes filicaulis. *Nigerian Journal of Health and Biomedical Sciences* 4(1): 30–34, Doi: 10.4314/njhbs.v4i1.11534.
64. Rios, M.Y., 2012. Natural alkamides: Pharmacology, chemistry and distribution. In: Vallisuta, O. and Olimat, S. M., editors. *Drug discovery research in pharmacognosy*. InTech, Rijeka. pp. 107–144.
65. Fabry, W., Okemo, P.O. and Ansorg, R., 1998. Antibacterial activity of East African medicinal plants. *Journal of Ethnopharmacology* 60(1): 79–84, Doi: 10.1016/s0378–8741(97)00128–1.

66. Sinei, K.A., et al., 2013. An investigation of the antimicrobial activity of Acmella cau-lirhiza. *African Journal of Pharmacology and Therapeutics* 2(4): 130–133.
67. Hadijja, N., 2017. *Antibacterial activity of Cochorus olitorius and Acmella caulirhiza on Streptococcus mutans, a cariogenic bacteria in dental caries.* Makerere University, Kampala.
68. Ngueguim, F.T., et al., 2012. Evaluation of Cameroonian plants towards experimen-tal bone regeneration. *Journal of Ethnopharmacology* 141: 331–337, Doi: 10.1016/j.jep.2012.02.041.
69. Cos, P., et al., 2002. Further evaluation of Rwandan medicinal plant extracts for their antimicrobial and antiviral activities. *Journal of Ethnopharmacology* 79: 155–163.
70. Donkeng Donfack, V.F., et al., 2014. Antimycobacterial activity of selected medici-nal plant extracts from Cameroon. *International Journal of Biological and Chemical Sciences* 8(1): 273–288, Doi: 10.4314/ijbcs.v8i1.24.
71. Fabry, W., Okemo, P. and Ansorg, R., 1996. Activity of east African medicinal plants against Helicobacter pylori. *Chemotherapy* 42(5): 315–317, Doi: 10.1159/000239460.
72. Clarkson, C., et al., 2004. In vitro antiplasmodial activity of medicinal plants native to or naturalised in South Africa. *Journal of Ethnopharmacology* 92(2–3): 177–191, Doi: 10.1016/j.jep.2004.02.011.
73. Gubler, D.J., 1998. Resurgent vector-borne diseases as a global health problem. *Emerging Infectious Diseases* 4(3): 442–450, Doi: 10.3201/eid0403.980326.
74. Ponlawat, A., et al., 2005. Insecticide susceptibility of Aedes aegypti and Aedes albop-ictus across Thailand. *Journal of Medical Entomology* 42(5): 821–825, Doi: 10.1603/0022–2585(2005)042[0821:ISOAAA]2.0.CO;2.
75. Saraf, D.K. and Dixit, V.K., 2002. Spilanthes acmella Murr.: Study on its extract spilan-thol as larvicidal compound. *Asian Journal of Experimental Biological Sciences* 16(1): 9–19.
76. Ramsewak, R.S., Erickson, A.J. and Nair, M.G., 1999. Bioactive N -isobutylamides from the flower buds of Spilanthes acmella. *Phytochemistry* 51: 729–732.
77. Sharma, A., et al., 2012. Insecticidal Toxicity of Spilanthol from Spilanthes acmella Murr. against Plutella xylostella L. *American Journal of Plant Sciences* 03(11): 1568–1572, Doi: 10.4236/ajps.2012.311189.
78. Prathibha, K.P., Raghavendra, B.S. and Vijayan, V.A., 2014. Larvicidal, ovicidal, and oviposition-deterrent activities of four plant extracts against three mosquito species. *Environmental Science and Pollution Research* 21(10): 6736–6743, Doi: 10.1007/s11356-014-2591-7.
79. Ottea, J.A., Payne, G.T. and Soderlund, D.M., 1990. Action of insecticidal N-Alkylamides at site 2 of the voltage-sensitive sodium channel. *Journal of Agricultural and Food Chemistry* 38(8): 1724–1728, Doi: 10.1021/jf00098a021.
80. Elliott, M., et al., 1986. Insecticidal amides with selective potency against a resis-tant (Super-Kdr) strain of houseflies Musca Domestica L. *Agricultural and Biological Chemistry* 50(5): 1347–1349, Doi: 10.1080/00021369.1986.10867574.
81. Ngueguim, T.F., et al., 2015. Acute and sub-acute toxicity of a lyophilised aqueous extract of the aerial part of Spilanthes africana Delile in rats. *Journal of Ethnopharmacology* 172: 145–154, Doi: 10.1016/j.jep.2015.06.035.
82. Ngogang, J., et al., 2015. Evaluation of the hypoglycaemic, hypolipidemic and antioxi-dant, properties of a Cameroonian polyherbal formulation on diabetic rats. *2nd Annual International Conference on Pharmaceutical Sciences*, Athens. pp. 45–46.
83. Campos-Cuevas, J.C., et al., 2008. Tissue culture of Arabidopsis thaliana explants reveals a stimulatory effect of alkamides on adventitious root forma-tion and nitric oxide accumulation. *Plant Science* 174(2): 165–173, Doi: 10.1016/j.plantsci.2007.11.003.

84. Ramírez-Chávez, E., Molina-Torres, J. and Ríos-Chávez, P., 2011. Natural distribution and alkamides production in Acmella radicans. Emirates *Journal of Food and Agriculture* 23(3): 275–282.

85. Uttaro, A.D., 2006. Biosynthesis of polyunsaturated fatty acids in lower eukaryotes. *IUBMB Life* 58(10): 563–571, Doi: 10.1080/15216540600920899.

86. Cahoon, E.B., et al., 2001. Formation of conjugated Δ8,Δ10-double bonds by Δ12-oleic-acid desaturase-related enzymes. Biosynthetic origin of calendic acid. *Journal of Biological Chemistry* 276(4): 2637–2643, Doi: 10.1074/jbc.M009188200.

87. Minto, R. and Blacklock, B., 2008. Biosynthesis and function of polyacetylenes and allied natural products. *Progress in Lipid Research* 47: 233–306, Doi: 10.1016/j.plipres.2008.02.02.

88. Greger, H., 1984. Alkamides: Structural relationships, distribution and biological activity. *Planta Medica* 50(5): 366–375, Doi: 10.1055/s-2007-969741.

89. Ikeda, Y., et al., 1984. Facile routes to natural acyclic polyenes synthesis of spilanthol and trail pheromone for termite. *Tetrahedron Letters* 25(45): 5117–5180.

90. Alonso, I.G., et al., 2018. A new approach for the total synthesis of spilanthol and analogue with improved anesthetic activity. *Tetrahedron* 74(38): 5192–5199, Doi: 10.1016/j.tet.2018.06.034.

91. Crombie, L., Krasinski, A.H.A. and Manzoor-I-Khuda, M., 1963. Amides of vegetable origin. Part X. The stereochemistry and synthesis of affinin. *Journal of the Chemical Society (Resumed)*: 4970–4976.

92. Wang, Z., et al., 1998. Efficient preparation of functionalized (E, Z) dienes using acetylene as the building block. *Journal of Organic Chemistry* 63(12): 3806–3807, Doi: 10.1021/jo980440x.

93. Tanaka, S., et al., 2012. Production method of (2E, 6Z, 8E)-N- isobutyl-2,6,8-decatrienamide (Spilanthol), and food or drink, fragrance or cosmetic, or pharmaceutical comprising the same. United States Patent, US 8,217,192 B2, 2012.

94. Tanaka, S., et al., 2012. Method for manufacturing Spilanthol and intermediate manufacturing product therefor. United States Patent, US 9,029,593 B2, 2012.

95. Mohamed, Y. and Hansen, T., 2013. Z-Stereoselective semi-reduction of alkynes: modification of the Boland protocol. *Tetrahedron* 69: 3872–3877.

96. Swamy, K.C.K., et al., 2018. Advances in chemoselective and/or stereoselective semi-hydrogenation of alkynes. Tetrahedron Letters 59(5): 419–429, Doi: 10.1016/j.tetlet.2017.12.057.

97. Alonso, I.G. and Pastre, J.C., 2015. A new approach for the total synthesis of spilanthol. 16th Brazilian Meeting on Organic Synthesis, Buzios.

98. Boonen, J., et al., 2010. Transdermal behaviour of the N-alkylamide spilanthol (affinin) from Spilanthes acmella (Compositae) extracts. *Journal of Ethnopharmacology* 127(1): 77–84, Doi: 10.1016/j.jep.2009.09.046.

99. Veryser, L., et al., 2014. N-alkylamides: from plant to brain. *Functional Foods in Health and Disease* 4(6): 264–275, Doi: 10.31989/ffhd.v4i6.6.

100. Artursson, P., Palm, K. and Luthman, K., 2001. Caco-2 monolayers in experimental and theoretical predictions of drug transport. *Advanced Drug Delivery Reviews* 46: 27–43, Doi: 10.1016/j.addr.2012.09.005.

101. Matthias, A., et al., 2004. Permeability studies of alkylamides and caffeic acid conjugates from Echinacea using a Caco-2 cell monolayer model. *Journal of Clinical Pharmacy and Therapeutics* 29(1): 7–13, Doi: 10.1046/j.1365-2710.2003.00530.x.

102. Boonen, J., et al., 2010. LC-MS profiling of N-alkylamides in Spilanthes acmella extract and the transmucosal behaviour of its main bio-active spilanthol. *Journal of Pharmaceutical and Biomedical Analysis* 53(3): 243–249, Doi: 10.1016/j.jpba.2010.02.010.

103. Woelkart, K., et al., 2009. Pharmacokinetics and tissue distribution of dodeca-2 E, 4 E, 8 E, 10 e / Z -tetraenoic acid isobutylamides after Oral administration in rats. *Planta Medica* 75(12): 1306–1313, Doi: 10.1055/s–0029–1185631.

104. Lambert, D.M. and Fowler, C.J., 2005. The endocannabinoid system: Drug targets, lead compounds, and potential therapeutic applications. *Journal of Medicinal Chemistry* 48(16): 5059–5087, Doi: 10.1021/jm058183t.

105. Gertsch, J., 2008. Immunomodulatory lipids in plants: Plant fatty acid amides and the human endocannabinoid system. *Planta Medica* 74(6): 638–650, Doi: 10.1055/s–2008–1034302.

106. Gong, J.P., et al., 2006. Cannabinoid CB2 receptors: Immunohistochemical localization in rat brain. *Brain Research* 1071(1): 10–23, Doi: 10.1016/j.brainres.2005.11.035.

107. Svizenska, I., Dubovy, P. and Sulcova, A., 2008. Cannabinoid receptors 1 and 2 (CB1 and CB2), their distribution, ligands and functional involvement in nervous system. *Pharmacology, Biochemistry and Behaviour* 2(90): 501–511, Doi: 10.1016/j.pbb.2008.05.010.

108. Pertwee, R., 1999. Pharmacology of Cannabinoid receptor ligands. *Current Medicinal Chemistry* 6(8): 365–664.

109. Quintans, J.S.S., et al., 2014. Natural products evaluated in neuropathic pain models - A systematic review. *Basic & Clinical Pharmacology & Toxicology* 114: 442–50, Doi: 10.1111/bcpt.12178.

110. Hajdu, Z., et al., 2014. Identification of endocannabinoid system-modulating N-alkylamides from Heliopsis helianthoides var. scabra and Lepidium meyenii. *Journal of Natural Products* 77(7): 1663–1669, Doi: 10.1021/np500292g.

111. Hohmann, J., et al., 2011. Alkamides and a neolignan from Echinacea purpurea roots and the interaction of alkamides with G-protein-coupled cannabinoid receptors. *Phytochemistry* 72(14–15): 1848–1853, Doi: 10.1016/j.phytochem.2011.06.008.

112. Faller, B., et al., 2006. Alkylamides from Echinacea are a new class of cannabino-mimetics - Cannabinoid type 2 receptor-dependent and -independent immunomodulatory effects. *Journal of Biological Chemistry*, 281(20): 14192–14206, Doi: 10.1074/jbc.M601074200.

113. Woelkart, K., et al., 2005. The endocannabinoid system as a target for alkamides from Echinacea angustifolia roots. *Planta Medica* 71(8): 701–705, Doi: 10.1055/s–2005–871290.

114. Chicca, A., et al., 2009. Synergistic immunomopharmacological effects of N-alkylamides in Echinacea purpurea herbal extracts. *International Immunopharmacology* 9(7–8): 850–858, Doi: 10.1016/j.intimp.2009.03.006.

115. García, M. del C., Adler-Graschinsky, E. and Celuch, S.M., 2009. Enhancement of the hypotensive effects of intrathecally injected endocannabinoids by the entourage compound palmitoylethanolamide. *European Journal of Pharmacology* 610: 75–80, Doi: 10.1016/j.ejphar.2009.03.021.

116. Howlett, A.C., et al., 2002. International union of pharmacology. XXVII. Classification of cannabinoid receptors. *Pharmacological Reviews* 54(2): 161–202, Doi: 10.1124/pr.54.2.161.

117. Cascio, M.G., et al., 2010. In vitro and in vivo pharmacological characterization of two novel selective cannabinoid CB2 receptor inverse agonists. Pharmacological Research 61(4): 349–354, Doi: 10.1016/j.phrs.2009.11.011.

8 Cameroonian *Allanblackia* Species

Jean Emmanuel Mbosso Teinkela
University of Douala
Université Libre de Bruxelles (ULB)

CONTENTS

DOI: 10.1201/9780429506734-8

8.1   INTRODUCTION

Allanblackia is a dioecious multipurpose tree genus of the family Clusiaceae found in the equatorial rainforests of West, East, and Central African regions extending from Tanzania to Sierra Leone. It has several uses including shade, timber, and medicine, but the main use is the production of edible oil from its seeds. The dry seeds contain about 67%–73% of solid white fat [1] and have been used traditionally for cooking and soap making. Recently, new uses of *Allanblackia* seed oil at an industrial scale for the manufacture of margarine and cosmetic products were reported, thus raising the international demand on a commercial scale. The moderately high melting point of the *Allanblackia* seed oil among other properties makes it superior to other alternative oils like palm oil. The oil has recently received the approval of the European Union (EU) Novel Food Regulations that certifies its safe usage in food products [2]. All members of the genus are apparently long-lived, have a long fruiting season, and are dioecious (they have separate male and female trees). *Allanblackia* trees are single stemmed, up to 40 m tall, with whorled branches, and bear the largest fruits of all plants in the African rainforest, particularly *Allanblackia stuhlmannii*. Its fruits can weigh up to 7 kg, depending on the species. The fruits drop when they are ripe, so plucking is not needed or recommended for producing [3]. With these data in mind, recent research efforts determined both the main chemical constituents and their phytopharmaceutical properties. Leaves, branches, fruits, roots, stem bark, and seed extracts as well as isolated components for some Cameroonian plants of the genus *Allanblackia*: *Allanblackia floribunda* Oliver, *Allanblackia monticola* Staner L.C., and *Allanblackia gabonensis* Pellegr. were explored, highlighting these plants as a natural source for potential pharmaceutical agents. A brief overview of these constituents as well as their bioactivities, antioxidant, antimicrobial, cytotoxic, antitumor, apoptotic, antiproliferative, anthelmintic, antileishmanial, hepato-nephroprotective, vasorelaxant, anticholinesterase, anti-noceptive, antiplasmodial, ejaculatory, analgesic and anti-inflammatory, is presented.

8.2   USES

A. floribunda is used traditionally to treat many ailments. In Cameroon, the decoction of the stem bark is used to treat dysentery or as gargle against toothache. The seeds are used in the manufacture of ointment against itching [4]. The extracts from leaves, stem bark, and roots are used alone or combined with other plants in several African countries such as Gabon, Congo, and Cameroon to treat respiratory infections, dysentery, diarrhea, and toothache [4] (Figure 8.1).

A. monticola has been used traditionally in Cameroon and other countries as a remedy against amoebic dysentery, diarrhoea, indigestion, pulmonary infections, skin diseases, headache, inflammation, and generalized pain [5] (Figure 8.2).

A. gabonensis is generally used in traditional medicine to improve virility in men and to treat infections such as dysenteries, colds, and toothaches [4,5]. People also use it to relieve pain, inflammation, and rheumatism (Figure 8.3).

FIGURE 8.1   Picture of *A. floribunda.*

FIGURE 8.2   Picture of *A. monticola.*

FIGURE 8.3   Picture of *A. gabonensis*.

8.3   ISOLATION TECHNIQUES USED

The extracts' residues are most often fractionated by flash column chromatography using silica gel (230–400 mesh) eluted with the mixture of cyclohexane–EtOAc, cyclohexane–EtOAc, EtOAc, and EtOAc–MeOH. The pure compounds were obtained either by direct crystallization or after further purification by column chromatography and crystallization [6].

8.4   PHYTOCHEMISTRY

8.4.1   SECONDARY METABOLITES FROM ALLANBLACKIA FLORIBUNDA

From the root bark extract of *A. Floribunda*, several phenolics have been isolated, namely 1,7-dihydroxyxanthone, macluraxanthone, volkensiflavone, morelloflavone, and 7'-O-glucoside of morelloflavone [6,7] (Figure 8.4).

From the stem bark of *A. floribunda*, several xanthones have been isolated named allanxanthone A, 1,5-dihydoxyxanthone, 1,5,6-trihydroxy-3,7-dimethoxyxanthone, and stigmasterol and stigmasterol-3-*O*-β-D-glucopyranoside [8] (Figure 8.5).

8.4.2   SECONDARY METABOLITES FROM ALLANBLACKIA MONTICOLA

From the stem bark of *Allanblackia monticola*, xanthone derivatives have been isolated: 3,6,7-trihydroxy-1-methoxyxanthone, 2,6-dihydroxy-1-methoxyxanthone, allanxanthone A, allanxanthone B, allanxanthone C tovophyllin A, rubraxanthone, and the compounds garciniafuran, epicathechin, norcowanin, α-mangostin, pentacyclic triterpene, oleanolic acid acetate, lupeol, betulinic acid and phytosterols, stigmasterol, and stigmasterol-3-*O*-β-D-glucopyranoside [9–12] (Figure 8.6).

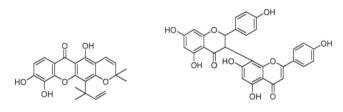

1,7-dihydroxyxanthone   morelloflavone   7'-O-glucoside of morelloflavone

macluraxanthone   volkensiflavone

FIGURE 8.4   Compounds isolated from the root bark of *Allanblackia floribunda*.

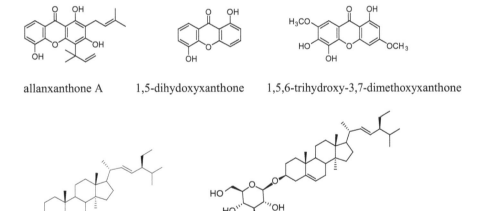

allanxanthone A   1,5-dihydoxyxanthone   1,5,6-trihydroxy-3,7-dimethoxyxanthone

Stigmasterol   stigmasterol-3-*O*-β-D-glucopyranoside

FIGURE 8.5   Compounds isolated from the stem bark of *Allanblackia floribunda*.

Chemical examination of the seeds led to the isolation and characterization of xanthone derivatives including allanxanthone E, 1,7-dihydroxy-3-methoxy-2-(3-methylbut-2-enyl)xanthone, α-mangostin, garciniafuran, allanxanthone C, and 1,6-dihydroxy-2,4-diprenylxanthone together with two pentacyclic triterpenes, friedelin and lupeol [13] (Figure 8.7).

The fruits yielded three benzophenones: garcinol, cambogin, and guttiferone F, along with one xanthone: allanxanthone A as active constituents [14] (Figure 8.8).

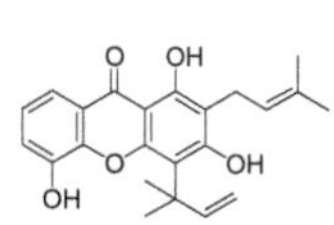

allanxanthone A 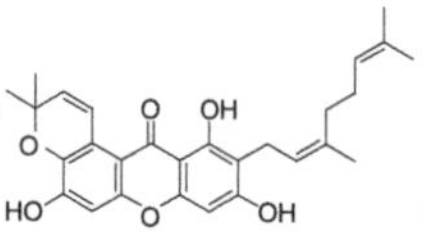　　allanxanthone B 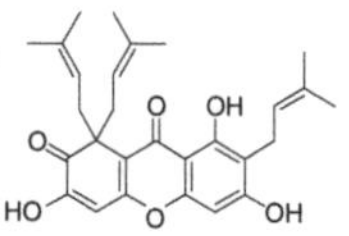　　allanxanthone C

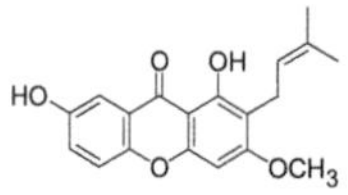

1,7-dihydroxy-3-methoxy-2- (3-methylbut-

2-enyl)xanthone

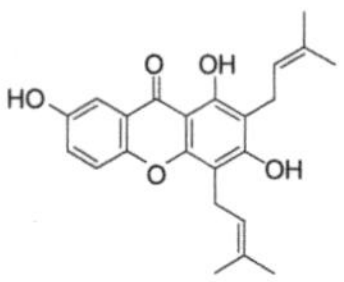

1,6-dihydroxy-2,4-diprenylxanthone

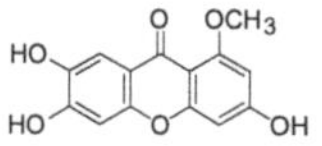

3,6,7-trihydroxy-1-methoxyxanthone 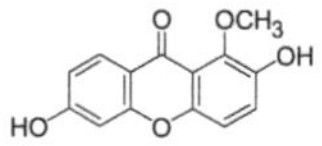　　2,6-dihydroxy-1-methoxyxanthone,

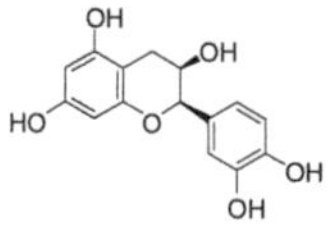

epicathechin 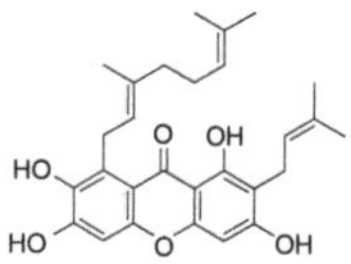　　norcowanin 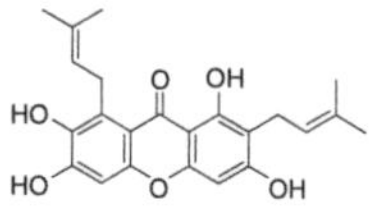　　α-mangostin

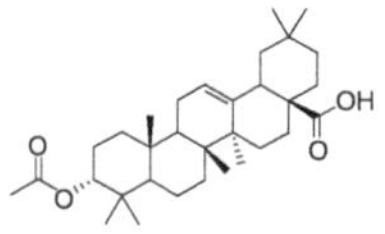

oleanolic acid acetate 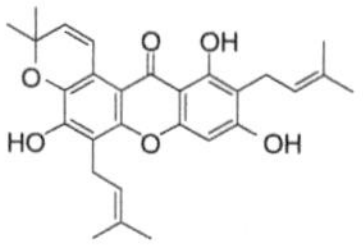　　tovophyllin A 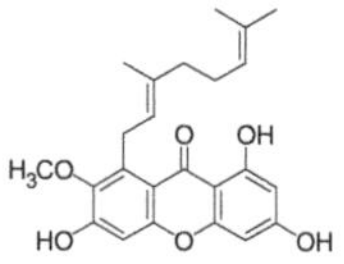　　rubraxanthone

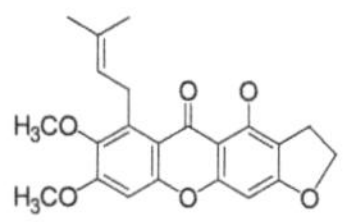

Garciniafuran 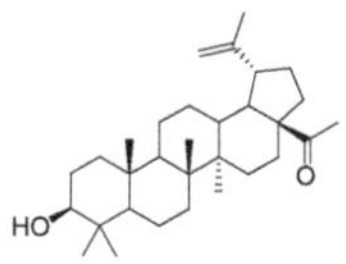　　betulinic acid 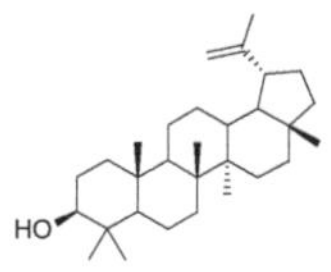　　lupeol

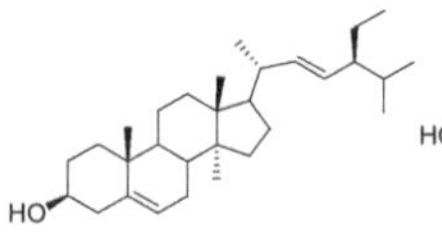

Stigmasterol 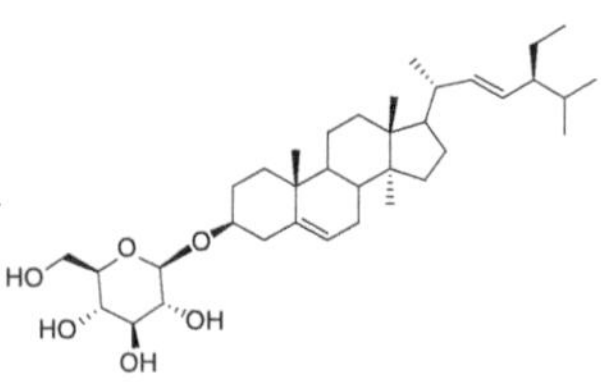　　stigmasterol-3-O-β-D-glucopyranoside

FIGURE 8.6 Compounds isolated from the stem bark of *Allanblackia monticola*.

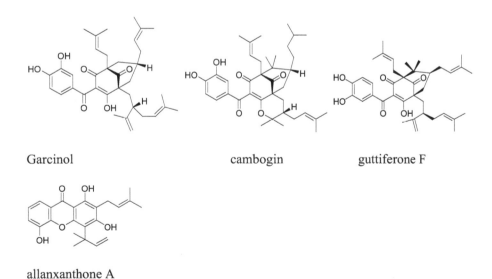

allanxanthone E   1,7-dihydroxy-3-methoxy-2-(3-methylbut-2-enyl)xanthone

α-mangostin   garciniafuran   allanxanthone C

1,6-dihydroxy-2,4-diprenylxanthone   friedelin   lupeol

FIGURE 8.7 Compounds isolated from the seeds of *Allanblackia monticola*.

Garcinol   cambogin   guttiferone F

allanxanthone A

FIGURE 8.8 Compounds isolated from the fruits of *Allanblackia monticola*.

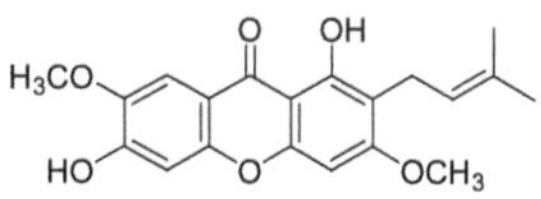

1,6-dihydroxy-3,7-dimethoxy-2   α-mangostin   tovophyllin A

-(3-methylbut-2-enyl)xanthone

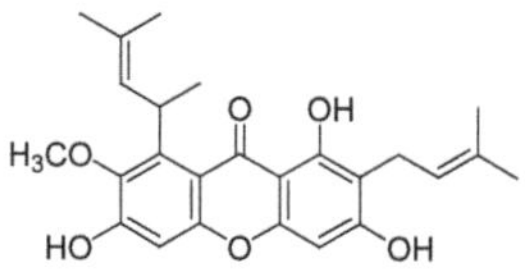

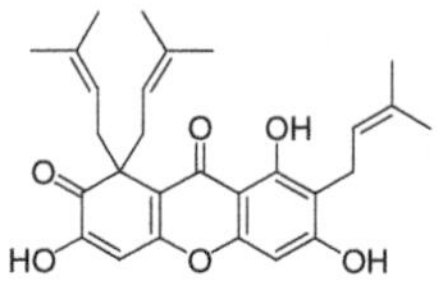

allanxanthone C 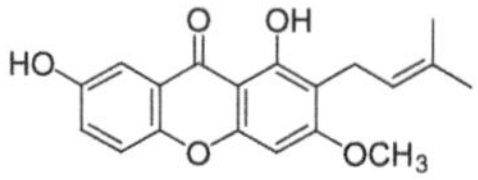   1,7-dihydroxy-3-methoxy-2   amentoflavone

-(3-methylbut-2-enyl)xanthone

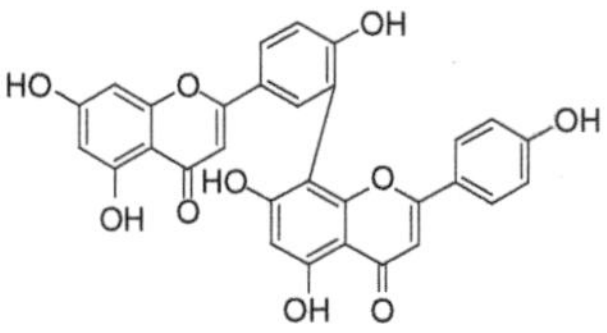

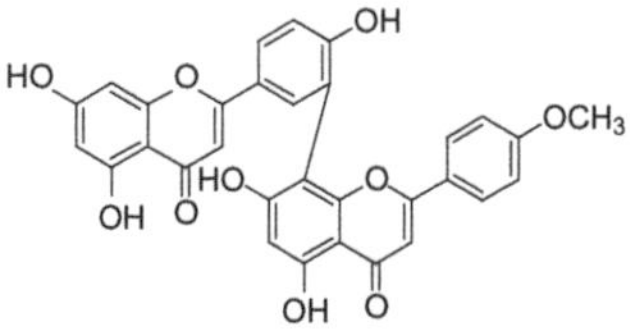

podocarpusflavone A 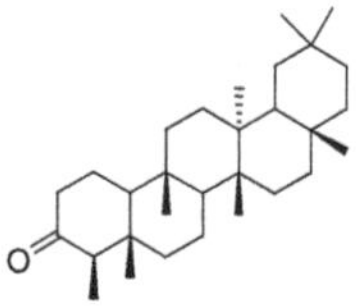   friedelan-3-one

FIGURE 8.9 Compounds isolated from the leaves of *Allanblackia monticola*.

Examination of the leaves yielded five prenylated xanthones [1,6-dihydroxy-3,7-dimethoxy-2-(3-methylbut-2-enyl)xanthone, α-mangostin, tovophyllin A, allanxanthone C, and 1,7-dihydroxy-3-methoxy-2-(3-methylbut-2-enyl)xanthone], two biflavonoid derivatives (amentoflavone and podocarpusflavone A), and one pentacyclic triterpene (friedelan-3-one) [15] (Figure 8.9).

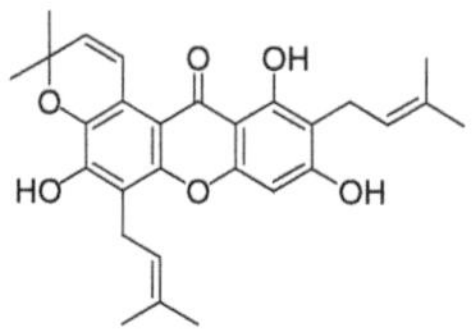

8.4.3 SECONDARY METABOLITES FROM *ALLANBLACKIA GABONENSIS*

The stem bark of *A. gabonensis* provided 11 compounds including seven xanthone derivatives allanxanthone D allanxanthone A, 1,5-dihydroxyxanthone, 1,7-dihydroxyxanthone, 1,3,6,7-tetrahydroxy-2-(3-methylbut-2-enyl)xanthone, forbexanthone, and 6-deoxyisojacareubin, one polyisoprenylated benzophenone guttiferone F, one flavanol epicathechin, and two phytosterols, β-sitosterol and campesterol [16] (Figure 8.10).

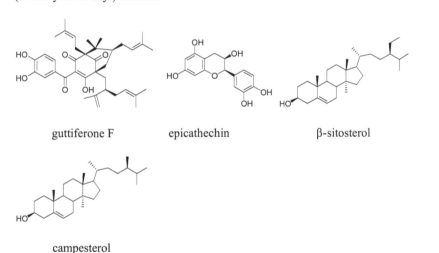

allanxanthone D   allanxanthone A   1,5-dihydroxyxanthone   1,7-dihydroxyxanthone

1,3,6,7-tetrahydroxy-2-   forbexanthone   6-deoxyisojacareubin

(3-methylbut-2-enyl)xanthone

guttiferone F   epicathechin   β-sitosterol

campesterol

FIGURE 8.10 Compounds isolated from the stem bark of *Allanblackia gabonensis*.

8.5 BIOLOGICAL ACTIVITIES OF SECONDARY METABOLITES AND EXTRACTS

8.5.1 *ALLANBLACKIA FLORIBUNDA*

8.5.1.1 Cytotoxic Activity

Three xanthone metabolites, allanxanthone A, 1,5-dihydoxyxanthone, and 1,5,6-tr ihydroxy-3,7-dimethoxyxanthone, isolated from *A. floribunda* were screened for *in vitro* cytotoxicity against the KB cancer cell line (derived from a nasopharyngeal carcinoma of human origin). They were moderately cytotoxic with EC_{50} values of 1.5 (allanxanthone A), 3.3 (1,5-dihydoxyxanthone), and 2.5 (1,5,6-trihydroxy-3,7-dimethoxyxanthone) mg/mL [8].

8.5.1.2   Antitumor Activity

1,7-dihydroxyxanthone, morelloflavone, and 7'-O-glucoside of morelloflavone and crude extract of the root bark of *A. floribunda* were evaluated for antitumor activity. The extract (42.46% at 100 µg/disc) exhibited only moderate tumor reducing activity; however, from the isolated individual compounds, morelloflavone and 7'-O-glucoside of morelloflavone, stronger activity was recorded with tumor inhibition percentages of 51.29% and 56.17%, respectively. 1,7-Dihydroxyxanthone with 13.89% activity was less active compared to morelloflavone and 7'-O-glucoside of morelloflavone. [7].

8.5.1.3   Antioxidant Activity

Free radical scavenging activity was determined by measuring the decrease in the visible absorbance of 2,2-diphenyl-1-picrylhydrazyl (DPPH) on addition of the plant methanolic extracts (leaves and fruits) of *A. floribunda*. Only fruit contained phlobatannins. IC_{50} values of 0.01, 0.02, and 0.1 mg/mL were recorded for vitamin E, leaves, and fruits, respectively. Total phenolic, total flavonoid, and proanthocyanidin contents were 65, 0.07, and 2.38 mg/g of powdered plant material for fruits, and 12, 51.35, 19.5 mg/g of powdered plant material for leaves as gallic acid, rutin, and catechin equivalents, respectively. Leaves are five times more potent as a free radical scavenger compared to the fruits though the fruit was found to contain a higher phenolic content [17]. 1,7-dihydroxyxanthone, morelloflavone, and 7'-O-glucoside of morelloflavone and crude extract of the root bark were evaluated for antioxidant activity. The DPPH radical scavenging test showed that at the concentration of 500 µg/mL, all the studied samples were able to scavenge more than 50% of the free DPPH radical. 7'-O-glucoside of morelloflavone showed the best activity, exhibiting 91.1% inhibition (IC_{50} of 49.1 µg/mL). This activity was not significantly ($p < 0.05$) different from that of ascorbic acid used as a reference antioxidant compound. The IC_{50} as determined by graphic extrapolation were 45.7, 49.1, 62.8, 76.3, and 488.5 µg/mL, respectively for vitamin C, morelloflavone, 7'-O-glucoside of morelloflavone, the crude extract, and 1,7-dihydroxyxanthone [7].

8.5.1.4   Antimicrobial Activity

1,7-dihydroxyxanthone, morelloflavone, 7'-O-glucoside of morelloflavone, and crude extract of the root bark were evaluated for antimicrobial activity. The antimycobacterial assays showed that the extract as well as the three compounds prevented the growth of *M. smegmatis* in the tested concentration range. Only the extract and morelloflavone were active on *M. tuberculosis*. The MIC values of 39.06 µg/mL for the extract and 19.53 µg/mL for morelloflavone were recorded on *M. smegmatis*. Results of the diffusion test demonstrated that the extract and morelloflavone prevented the growth of all the tested organisms. Morelloflavone and 7'-O-glucoside of morelloflavone were active on 11 of the 18 (61.1%) studied organisms including Gram-positive and Gram-negative bacteria and fungi. The results of MIC determinations indicated values ranging from 19.53 to 312.50 µg/mL for the extract on most of the tested microorganisms. As previously observed, 7'-O-glucoside of morelloflavone was selectively active. The MIC value of 9.76 µg/mL for the extract was recorded against *E. aerogenes*. The MIC of 4.88 µg/mL was noted with morelloflavone on *Trichophyton rubrum*. The reference antibiotics exhibited MICs ranging from 2.44 to 19.53 µg/mL. The inhibition potentials of the extract and morelloflavone can be considered important when regarding the antibacterial and antifungal activities of the reference antibiotic: nystatin on *T. rubrum* [7].

Four extracts (aqueous, ethanolic, methanolic, and 0.1 N HCl) of *A. floribunda* seeds were evaluated for their antioxidant properties and scavenging activities. The FRAP antioxidant capacities (mg equivalent of ascorbic acid/g d.m) showed that extracts had higher antioxidant properties, highest in the ethanolic fraction with values 21.80 ± 0.50 (ethanolic), 15.82 ± 0.48 (methanolic), 11.02 ± 0.08 (aqueous), and 3.72 ± 0.04 (0.1 N HCl). With ABTS (acide 2,2'-azino-bis(3-éthylbenzothiazoline-6-sulphonique), the inhibition percentage fluctuated between $78.9\% \pm 0.98\%$ and $80.2\% \pm 0.05\%$ [18].

8.5.1.5 Ejaculatory Activities

Aqueous and ethanol extracts (2.5, 10, 20, 40, or 60 mg/kg) from the stem bark of *A. floribunda* were evaluated for their ejaculatory activities in rats in the presence or absence of dopamine (60 mg/kg). Furthermore, electromyographic activities of the bulbospongiosus muscles were recorded in five groups of spinal rats pretreated orally during 8 d with extracts (150 and 300 mg/kg) in the presence of dopamine. Sequential treatments of rats with extracts significantly decreased the occurrence of ejaculation induced by dopamine up to 88.94% inhibition. The oral pretreatment with both extracts significantly decreased the ejaculation induced by dopamine with the highest inhibition of 89.79%. The two extracts had inhibitory activities on ejaculation. The inhibitory effect of these extracts on fictive ejaculation in rats may be directly mediated through dopaminergic pathways. Inhibition of ejaculation caused by these extracts could support its use in patients suffering from rapid ejaculation [19].

8.5.1.6 Antimalarial Activity

The linear regression allowed determining the IC_{50} of the tested compounds (1,7-dihydroxyxanthone, macluraxanthone, morelloflavone, volkensiflavone, and morelloflavone 7-*O*-glucoside) isolated from *A. floribunda* stem bark. After 24 h of contact with the parasite, volkensiflavone (IC_{50}: 0.99 µg/mL) and macluraxanthone (IC_{50}: 0.46 µg/mL) displayed the best activity on the F_{32} strain, while chloroquine (IC_{50}: 0.036 µg/mL) was used as a reference. With FcM_{29} strain, volkensiflavone (IC_{50}: 0.93 µg/mL) and macluraxanthone (IC_{50}:0.33µg/mL) remained the most active compounds, but the volkensiflavone was more active than the reference (chloroquine: IC_{50}: 0.57 µg/mL). After 72 h of contact, macluraxanthone (IC_{50}: 0.36 µg/mL) exhibited high activity on the F_{32} strain, and with FcM_{29} strain, it was most active (IC_{50}: 0.27 µg/mL). Additionally, several other compounds also showed good activities: volkensiflavone (IC_{50}: 0.95 µg/mL) and α-mangostin (IC_{50}: 0.33 µg/mL). For the most active compounds (macluraxanthone), we noted that when it was allowed to go from 24 to 72 h of contact with parasite, there was increased activity of 0.1 µg/mL with the F_{32} strain and 0.06 µg/mL with the FcM_{29} strain [6].

8.5.2 ALLANBLACKIA MONTICOLA

8.5.2.1 Antibacterial and Antifungal Activities

Crude extracts of the stem bark of *A. monticola*, allanxanthone B, rubraxan-thone, tovophyllin A, garciniafuran, allanxanthone C, norcowanin, and mangos-tin were tested for their antimicrobial potency against the Gram-positive bacteria *Staphylococcus aureus* (ATCC 6538), Gram-negative *Vibrio anguillarum* (ATCC 19264), and the pathogenic fungi *Candida tropicalis* (ATCC 66029) in an agar well diffusion assay. Crude extract and the seven xanthones were found to be inactive against *V. angillarium* and *C. tropicalis*. While against *S. aureus*, crude extracts and rubraxanthone displayed moderate activity, mangostin also displayed moderate activity; compounds (allanxanthone B, tovophyllin A, garciniafuran, allanxanthone C, and norcowanin) were found to be inactive. Mangostin and crude extract are half as active against *S. aureus* than the reference compound oxacillin [9,11].

8.5.2.2 Antiplasmodial and Cytotoxicity Activities

The crude methanol extract of stem bark, allanxanthone C, norcowanin, allaxan-thone B, and α-mangostin were assayed for their antiplasmodial activities against two strains of *Plasmodium falciparum*: F_{32} (chloroquine-sensitive) and FcM_{29} (chloro-quine resistant) and for their cytotoxicity against human melanoma cells (A375). All these compounds were moderately active against the two strains of *P. Falciparum*. IC_{50} values obtained for allanxanthone C and norcowanin ranged from 0.6 to 8.9 mg/mL and for allaxanthone B 3.1 to 3.9 μg/mL with F_{32} and FcM_{29} layers. Their cytotoxicity was estimated on human melanoma cells (A375), and the cytotoxicity/antiplasmodial ratio ranged between 15.45 and 30.46 [6,11].

8.5.2.3 Anti-Inflammatory and Anti-Nociceptive Activities

The anti-inflammatory activity "*in vivo*" of the methylene chloride/methanol extract, methanol, and methylene chloride fractions of stem barks of *Allanblackia monticola*, administered orally at doses of 37.5; 75; 150 and 300 mg/kg, was evaluated on carra-geenan-induced edema in rats to determine the most active fraction. Indomethacin, inhibitor of cyclo-oxygenase, was used as a reference drug. The effects of the most active fraction were then examined on the rat paw edema caused by histamine, sero-tonin, arachidonic acid, and dextran followed by its ulcerogenic effect. The methylene chloride fraction was more effective on the edema caused by the carrageenan. The anti-nociceptive activity of the methylene chloride fraction was assessed using the acetic acid-induced abdominal constriction model, formalin test, and hot plate test. At 150 mg/kg, *A. monticola* caused maximum inhibition of inflammation induced by carrageenan (83.33%), by histamine (42.10%), by dextran (40.29%), and by arachi-donic acid (64.28%). *A. monticola* (75–300 mg/kg) did not cause significant modifi-cation of the edema induced by serotonin. Concerning the anti-nociceptive properties of the plant, the methylene chloride fraction (75–300 mg/kg) caused a dose-dependent inhibition on abdominal contractions induced by acetic acid (32.34%–77.37%) and significantly inhibited the inflammatory pain caused by formalin (40.71%–64.78%). *A. monticola* did not increase the latency time in the hot plate test. Like indometha-cin (10 mg/kg), the fraction at the dose of 150 mg/kg caused ulceration of the gastric mucous membrane in treated rats. Thus, *A. monticola* has anti-inflammatory and analgesic activities with gastric ulcerative side effects [20].

8.5.2.4   Leishmanicidal and Anticholinesterase Activities

In a preliminary antiprotozoal screening of several Clusiaceae species, the methanolic extracts of *Allanblackia monticola* fruits showed high *in vitro* leishmanicidal activity. Further bioguided phytochemical investigation led to the isolation of guttiferone F, which showed strong leishmanicidal activity *in vitro*, with an IC_{50} value of 0.16 μM, comparable to that of the reference compound, miltefosine (0.46 μM). Although the leishmanicidal activity is promising, the cytotoxicity profile of these compounds prevents at this state further *in vivo* biological evaluation. In addition, all the isolated compounds (garcinol, cambogin, guttiferone F, and allanxanthone A) were tested *in vitro* for their anticholinesterase properties. The three benzophenones showed potent anticholinesterase properties toward acetylcholinesterase (AChE) and butyrylcholinesterase (BChE). For AChE, the IC_{50} value (0.66 μM) of garcinol was almost equal to that of the reference compound galantamine (0.50 μM). Furthermore, guttiferone F with an IC_{50} value of 3.50 μM was more active than galantamine ($IC_{50} = 8.5$) against BChE [14].

8.5.2.5   Apoptotic and Antiproliferative Activities

Three selected xanthones (allanxanthone E, 1,7-dihydroxy-3-methoxy-2-(3-methylbut-2-enyl)xanthone, and α-mangostin) were evaluated for their apoptotic and antiproliferative activities against human leukemic B lymphocytes, such as the hairy cell leukemia-derived ESKOL cell line and cells from B-CLL (B-cell chronic lymphocytic leukemia) patients. They display proapoptotic properties on cells from B-CLL patients, with an activity slightly lower than but comparable to that of flavopiridol. Furthermore, 1,7-dihydroxy-3-methoxy-2-(3-methylbut-2-enyl) xanthone and α-mangostin exhibit potent cytotoxic activity on a B-cell line derived from a hairy cell leukemia patient. Interestingly, while α-mangostin is also somewhat cytotoxic for normal B lymphocytes, allanxanthone E and 1,7-dihydroxy-3-methoxy-2-(3-methylbut-2-enyl)xanthone do not alter the viability of these cells, even at the highest concentrations tested (23.0, 31.0, and 24.0 μM for compounds allanxanthone E, 1,7-dihydroxy-3-methoxy-2-(3-methylbut-2-enyl)xanthone, and α-mangostin, respectively). These compounds, especially 1,7-dihydroxy-3-methoxy-2-(3-methylbut-2-enyl)xanthone, thus, exhibit selective toxicity and proapoptotic activity on tumor cells from B-CLL patients and warrant further studies as potentially interesting molecules for the therapy of this leukemia [13].

8.5.2.6   Antioxidant and Anti-Inflammatory Activities

α-Mangostin, lupeol, and betulinic acid were screened for antioxidant activity using a free radical scavenging method. These isolated compounds were further tested for anti-inflammatory properties using carrageenan-induced model. Methylene chloride fractions showed concentration-dependent radical scavenging activity, by inhibiting 1,1-diphenyl-1-picryl-hydrazyl radical (DPPH) with an IC_{50} value of 14.60 μg/mL. α-Mangostin and betulinic acid (500 μg/mL) showed weak radical scavenging activity with a maximum inhibition reaching 38.07 and 26.38 μg/mL, respectively. Betulinic acid, lupeol, and α-mangostin (5 and 9.37 mg/kg) showed anti-inflammatory activity with a maximum inhibition of 57.89%, 57.14%, and 38.70%, respectively. Methylene chloride fractions of *Allanblackia monticola* and some derivatives have antioxidant and anti-inflammatory activities [12].

8.5.2.7   Antimalarial and Vasorelaxant Activities

Four xanthones (α -mangostin, tovophyllin A, allanxanthone C, and 1,7-dihydroxy-3-methoxy-2-(3-methylbut-2-enyl)xanthone), two biflavonoid derivatives (amentoflavone 6 and podocarpusflavone A) and one pentacyclic triterpene (friedelan-3-one) and a crude methanolic extract of *A. monticola* leaves were each tested for antimalarial activity *in vitro*, using the chloroquine-sensitive F32 and chloroquine-resistant FcM29 strains of *Plasmodium falciparum*; the median inhibitory concentrations (IC$_{50}$) recorded varied from 0.7 to 83.5 mg/mL. The cytotoxicity of the compounds and crude extract, against cultures of human melanoma cells (A375), were then investigated, and cytotoxicity/antimalarial IC$_{50}$ ratios of 0.6–16.75 were recorded. In tests involving aortic rings from guinea pigs, a crude extract of the leaves of *A. monticola* was found to induce concentration-dependent vasorelaxation, causing up to 82% and 42% inhibition of noradrenaline- and KCl-induced contractions, respectively. The corresponding values for α-mangostin and amentoflavone when tested against noradrenaline-induced contractions were approximately 18% and 35%, respectively [15].

8.5.3   *ALLANBLACKIA GABONENSIS*

8.5.3.1   Antimicrobial Activity

Allanxanthone D, allanxanthone A, and forbexanthone isolated from the stem bark of *A. gabonensis* were evaluated for antimicrobial activities against a range of Gram-negative and Gram-positive bacteria. They exhibited a wide spectrum of antimicrobial activity with important inhibition either on Gram-positive and Gram-negative bacteria, yeasts, and mycelial fungus. The MIC values range from 1.22 to 39.06 mm/mL (3.72–119.1 mm) on Gram-negative bacteria, 1.22 to 4.88 m/mL (3.72–14.88 mm) on Gram-positive bacteria, and 1.22 to 19.53 (3.72–59.54 mm/mL) on Candida species, while the MIC obtained on the mycelial fungi was 1.22 mg/mL (3.72 mm/mL). The activity of allanxanthone D is greater than that of reference antibiotics with an MIC value of 1.22 mg/mL on Gram-positive (*P. vulgaris*, *P. aeruginosa*) and some of the tested Gram-negative bacteria (*S. faecalis*, *S. aureus*, and *B. cereus*) as well as against *Candida glabrata* and *Absidia* sp. Xanthones are known to complex irreversibly with nucleophilic amino acids in proteins, often leading to the inactivation of proteins and loss of function. This appeared to be the possible mechanism by which allanxanthone D, allanxanthone A and forbexanthone exhibited their antimicrobial effects [16].

8.5.3.2   Antileishmanial Activity

Allanxanthone D, 1,7-dihydroxyxanthone, 1,3,6,7-tetrahydroxy-2-(3-methylbut-2-en yl)xanthone, and epicathechin isolated from the stem bark of *A. gabonensis* were evaluated for their activity against *Leishmania amazonensis in vitro*. Xanthones

(1,3,6,7-tetrahydroxy-2-(3-methylbut-2-enyl) xanthone and allanxanthone D) have shown to be more active with IC_{50} of 13.3 and 13.9 mg/mL, respectively, than flavonoids (epicathechin). Among xanthones, tetraoxygenated prenylated xanthones were more active than dioxygenated simple xanthones, and their activities may be attributed to the presence of a prenyl group at position 2 or 4 [16].

8.5.3.3 Analgesic Activity

The analgesic effect of aqueous extract of the stem bark has been evaluated using acetic acid, formalin, hot-plate test, tail immersion, and paw-pressure test. It showed significant protective effects against chemical stimuli (acetic acid and formalin) in the mouse. Administered orally at the doses of 100–400 mg/kg, it exhibited a protective effect of at least 69.78% on the pain induced by acetic acid and also reduced first (67.18% at 200 mg/kg) and second (83.87% at 400 mg/kg) phase of pain-induced formalin. It also produced a significant increase of the threshold of sensitivity to pressure and hot plate-induced pain in the rats. These suggest peripheral and central analgesic activities [21].

8.5.3.4 Anti-Inflammatory Activity

The anti-inflammatory effect of an aqueous extract of the stem bark was also investigated on carrageenan, histamine or serotonin induced by paw edema. The aqueous extract of stem bark of *A. gabonensis* administrated p.o. showed significant activity against paw edema induced by carrageenan, with a maximum percentage of inhibition reaching the 74.01% at the preventive test at a dose of 200 mg/kg. The aqueous extract exhibited a significant reduction of paw edema induced by both histamine and serotonin with a maximal inhibition of 56.94% (200 mg/kg) and 40.83% (100 mg/kg), respectively. These suggest an anti-inflammatory effect of the stem bark of *A. gabonensis* [21].

8.5.3.5 Hepato-Nephroprotective Activity

The administration of aqueous extract of the stem bark of *A. gabonensis* at the dose of 100 and 200 mg/kg induces significant decreases in the activities of serum SGPT (serum glutamic pyruvic transaminase), SGOT (serum glutamic oxaloacetic transaminase), and serum bilirubin levels compared with the APAP-treated group. The variation of the serum creatinine level was not statistically significant compared with the APAP-treated group. These results were observed as well in prevention as well as curative treatments. *A. gabonensis* has produced significant hepato-nephroprotective activity by reducing the serum effect of MDA [22].

8.5.3.6 Antioxidant Activity

It was observed that the administration of aqueous extract of the stem bark of *A. gabonensis* at the dose of 100 and 200 mg/kg decreases liver and kidney CAT ($p < 0.05$), SOD ($p < 0.01$), and GSH ($p < 0.01$) level in N-acetyl-para-aminophenol (APAP) group during preventive studies ($p < 0.05$) as compared to the control animals. No significant variation was found for CAT during curative studies in both liver and kidney. However, a significant increase of CAT was observed in the APAP + 100 mg/kg extract group compared to the APAP-toxicity group ($p < 0.05$). A significant increase was observed for SOD in the APAP + 200 mg/kg extract group in the liver during the preventive test, while this increase was observed in both the APAP + 100 and 200 mg/kg extract groups during the curative test. Silymarin used as reference substance significantly corrected the decrease of SOD level induced by APAP. A significant increase of GSH level was observed in the serum of the APAP + 200 mg/kg extract group as compared to the APAP group during prevention and curative studies (p < 0.01). MDA significantly produced an increase in enzymatic antioxidant activities (SOD and CAT) and non-enzymatic antioxidant (GSH) levels. *A. gabonensis* also showed a significant decrease in transaminase, bilirubin, and creatinine in APAP intoxicated rats at the doses of 100 and 200 mg/kg [22].

8.6 CONCLUSION

Allanblackia is a dioecious multipurpose tree genus (family Clusiaceae) occurring in the equatorial rainforests of Africa extending from Tanzania to Sierra Leone. Besides its traditional main use for producing edible oil from its seeds, local communities use *Allanblackia* species as medicine and timber. Taking into consideration the biological activities of phytochemicals found, *Allanblackia* represents a new source of chemically relevant constituents for pharmacological purpose. Throughout this review, we highlighted the recent progress performed in the extraction, isolation, and identification of molecules from stem bark, leaves, branches, seeds, fruits, roots, and wood of *Allanblackia floribunda*, *Allanblackia gabonensis*, and *Allanblackia monticola* collected in Cameroon. Further investigations focused on the phytopharmacological properties of resulting plant extracts and isolated molecules, and in some cases, their performance with reference drugs is compared. Hence, antioxidant, antimicrobial, cytotoxic, antitumor, apoptotic, antiproliferative, anthelmintic, antileishmanial, hepato-nephroprotective, vasorelaxant, anticholinesterase, anti-noceptive, antiplasmodial, ejaculatory, analgesic, and anti-inflammatory properties were evaluated, emphasizing a wide spectrum of biological activities. These preliminary results may serve for an in-depth study and identification of the most active substances, following a modern drug development program.

REFERENCES

1. Sefah, W., 2006. Extraction and characterisation of vegetable fat from Allanblackia floribunda. Thesis submitted to the department of Biochemistry and Biotechnology in partial fulfilment of the requirement for the award of the Master of Science (M.Sc.) degree in food science and technology. Kwame Nkumah University of Science and Technology, p 146.
2. Hermann, M., 2009. The impact of the European Novel Food Regulation on trade and food innovation based on traditional plant foods from developing countries. *Food Policy* 34: 499–507.
3. The Allanblackia-tree [On line], [17/11/2020]. Available on: https://www.allanblacki-apartners.org/content/allanblackia-tree.
4. Vivien, J., Faure, J.J., 1979. *Arbre des forêts d'Afrique Centrale. Espèce du Cameroun.* Agence de Coopération Culturelle et Technique, Paris. pp. 212–213.

5. Raponda-Waker, A., Sillans, R., 1961. *Les plantes utiles du Gabon*. Paul Le Chevalier, Paris, pp. 216–219.

6. Azebaze, A.G.B, Mbosso, T.J.E., Nguemfo, E.L., Valentin, A, Dongmo, A.B., Vardamides, J.C., 2015. Antiplasmodial activity of some phenolic compounds from Cameroonians *Allanblackia*. *African Health Sciences* 15(3): 835–840.

7. Kuete, V., Azebaze, A.G.B., Mbaveng, A.T., Nguemfo, E.L., Tshikalange, E.T., Chalard, P., Nkengfack, A.E., 2011. Antioxidant, antitumor and antimicrobial activities of the crude extract and compounds of the root bark of *Allanblackia floribunda*. *Pharmaceutical Biology* 49(1): 57–65.

8. Nkengfack, A.E., Azebaze, A.G.B., Vardamides, J.C., Fomum, Z.T., Van Heerden, F.R., 2002. A prenylated xanthone from *Allanblackia floribunda*. *Phytochemistry* 60(4): 381–384.

9. Azebaze, A.G.B., Meyer, M., Bodo, B., Nkengfack, E.A., 2004. Allanxanthone B, a polyisoprenylated xanthone from the stem bark of *Allanblackia monticola* Staner L.C. *Phytochemistry* 65(18): 2561–2564.

10. Ngouela, S., Zelefack, F., Lenta, N.B., Ngouamegne, T.E., Tchamo, N.D., Tsamo, E., Connolly, D.J., 2005. Xanthones and other constituents of *Allanblackia monticola* (Guttiferae). *Natural Product Research* 19(7): 685–688.

11. Azebaze, A.G.B., Meyer, M., Valentin, A., Nguemfo, E.L., Fomum, Z.T., Nkengfack, A.E., 2006. Prenylated xanthone derivatives with antiplasmodial activity from *Allanblackia monticola* Staner L.C. *Chemical and Pharmaceutical Bulletin* 54(1): 111–113.

12. Nguemfo, E.L., Dimo, T., Dongmo, A.B., Azebaze, A.G.B., Alaoui, K., Asongalem, A.E., Cherrah, Y.P. Kamtchouing, P., 2009. Anti-oxidative and anti-inflammatory activities of some isolated constituents from the stem bark of *Allanblackia monticola* Staner L.C (Guttiferae). *Inflammopharmacology* 17(1): 37–41.

13. Azebaze, A.G.B., Menasria, F., Meyer, M., Noumi, L.G.N., Nguemfo, E.L., Nkengfack, A.E., Kolb, J.P., 2009. Xanthones from the seeds of *Allanblackia monticola* and their apoptotic and antiproliferative activities. *Planta Medica* 75(3): 243–248.

14. Lenta, N.B., Vonthron-Sénécheau, C., Weniger, B., Devkota, P.K., Ngoupayo, J., Kaiser, M., Naz, Q., Choudhary, I.M., Tsamo, E., Sewald, N., 2007. Leishmanicidal and cholinesterase inhibiting activities of phenolic compounds from *Allanblackia monticola* and *Symphonia globulifera*. *Molecules* 12(8): 1548–1557.

15. Azebaze, A.G.B., Dongmo, A.B., Meyer, M., Ouahouo, B.M.W., Valentin, A., Nguemfo, E.L., Nkengfack, A.E., Vierling, W., 2007. Antimalarial and vasorelaxant constituents of the leaves of *Allanblackia monticola* (Guttiferae). *Annals of Tropical Medicine & Parasitology* 101(1): 23–30.

16. Azebaze, A.G.B., Ouahouo, M.W.B., Valentin, A., Kuete, V., Nguemfo, E.L., Acebey, L., Beng, V.P., Nkengfack, A.E., Meyer, M., 2008. Antimicrobial and antileishmanial xanthones from the stem bark of *Allanblackia gabonensis* (Guttiferae). *Natural Product Research* 22(4): 333–341.

17. Ayoola, A.G., Ipav, S.S., Sofidiya, O.M., Adepoju-Bello, A.A., Coker, A.B.H., Odugbemi, O.T., 2008. Phytochemical Screening and Free Radical Scavenging Activities of the Fruits and Leaves of *Allanblackia floribunda* Oliv (Guttiferae). *International Journal of Health Research* 1(2): 87–93.

18. Boudjeko, T., Ngomoyogoli, K.E.J., Woguia, L.A., Yanou, N.N., 2013. Partial characterization, antioxidative properties and hypolipidemic effects of oilseed cake of *Allanblackia floribunda* and *Jatropha curcas*. *BMC Complementary Medicine and Therapies* 13: 352.

19. Sanda, K.A., Miegueu, P., Bilanda, D.C., Faleu, N.N.M., Watcho, P., Djomeni, P.D., Kamtchouing, P., 2013. Ejaculatory activities of *Allanblackia floribunda* stem bark in spinal male rats. *Pharmaceutical Biology* 51(8): 1014–1020.

20. Nguemfo, E.L., Dimo, T., Azebaze, A.G.B., Asongalem, A.E., Alaoui, K., Dongmo, B.A., Cherrah, Y., Kamtchouing, P., 2007. Anti-inflammatory and anti-nociceptive activities of the stem bark extracts from *Allanblackia monticola* Staner L.C. (Guttiferae). *Journal of Ethnopharmacology* 114(3): 417–424.
21. Ymele, V.E., Dongmo, A.B., Dimo, T., 2013. Analgesic and anti-inflammatory effect of aqueous extract of the stem bark of *Allanblackia gabonensis* (Guttiferae). *Inflammopharmacology* 21(1): 21–30.
22. Ymele, V.E., Donfack, M.F., Temdie, J.R., Ngueguim, T.F., Donfack, H.J., Dzeufiet, D.D., Dongmo, B.A., Dimo, T., 2012. Hepatho-nephroprotective and antioxidant effect of stem bark of *Allanblackia gabonensis* aqueous extract against acetaminophen-induced liver and kidney disorders in rats. *Journal of Experimental and Integrative Medicine* 2(4): 337–344.

Index

Note: **Bold** page numbers refer to tables and *italic* page numbers refer to figures.